How to Make a Universe

And Planet Life

Chris Beatty

Although the author has made every effort to ensure
that all information is correct, this is the universe and
life on Earth that we are talking about, so it should be
allowed to get a few things wrong.

This book contains short fictional stories. Any
resemblance to actual persons, living or dead, or
actual builders of universes and life on our planet, or
actual other life-supporting planets and their
equipment is purely coincidental.

universeplanetlife@gmail.com

To all who have gone before from whom
we have learned so much.

How to Make a Universe
And Planet Life

Chris Beatty is an engineer who lives in Berkshire, England. He specialises in 'Product Lifecycle Management'. This is his first published book.

Foreword

This is a genuine attempt to understand how to make a universe and how to create life on our planet. How can something so big be covered in just a short book? Because we are looking at principles rather than details.

It is down to my background as an engineer. I am experienced at going into factories to work out how to make their product development more efficient. This is done by cutting through what is on the surface to understand what is going on underneath.

With creation, we know that it will already be done in the best and simplest way; it is a question of working out what this is. Maybe this approach has uncovered some new insights.

However, the main aim is for you, dear reader, to take a little time to think about things that can seem impenetrable and the domain of experts but are fundamental to our own existence.

The book is written in a light-hearted way. It is not meant to be a serious scientific study, but inevitably we must be prepared to think about complex science.

I hope you will enjoy it and find it thought-provoking.

Chris Beatty

Contents

Introduction

Have you thought how the things we create are often based on a simple concept or idea?

James Dyson saw paint being separated in an air extractor using an industrial cyclone and thought, I could apply the same principle to a domestic vacuum cleaner. With this idea an extremely successful business was created.

We have a water softener. The underlying concept is 'ion exchange' in which the ions that make the water hard – calcium and magnesium – are exchanged with ions that make it soft – sodium ions from the block salt. Most of us have little idea what this really means and still less how to make that work, but with this simple underlying concept we can luxuriate in the bath, and when we open the toilet cistern it does not look like a recently discovered cave.

Concepts must be turned into designs, and *the best design is the simplest one that works*. Some of you will know the Screwpull wine bottle cork remover. There have been many designs of corkscrew, but this must be one of the simplest. There is the plastic part that you hold over the top of the bottle, and the bit with the corkscrew and handle that you wind into the cork. As you keep turning the cork pops out. A simple design that works.

When we look at the creation of the universe and of life on Earth, we see the same thing, simple underlying concepts and designs. This is significant for all of us who are not Albert Einstein or Stephen Hawking. It means that we can understand these things at a conceptual level.

Let us think about life on Earth, what are the underlying concepts and designs?

The most fundamental concept is that Life is just one thing. It may seem to us that there are many different types of life, such as trees and fish and flowers and birds, but these are different forms of one thing called Life. It must be like this for everything to be in balance.

Of course, in the beginning there wasn't all the diversity that we see today, Life started simple and became more complex and varied. There is a clever idea to achieve this; everything is made of cells. The simplest organism can be a single cell, but the same design can be used for making complex organisms with many cells. Right from the start there was a design that could go from simple to more complex and diverse.

To be alive, an organism must do something. The most fundamental thing it does is take in nutrients to grow and then reproduce. Even a single-celled organism can do this. Then there is DNA, another clever design concept. This code tells an organism how to build itself and then how to function. However, if organisms just created clones of themselves there would never be the diversity, so there is also a mechanism through which the code can be changed to build different and more complex organisms.

This is done from one generation to the next, so individual organisms are born, grow up, go along a bit, and then die, whilst Life itself goes on living and becoming more sophisticated. We see the same pattern in ourselves; individual cells in our body keep dying and being replaced while we go on living.

How about the universe, what are its underlying concepts and designs?

The most remarkable thing is that the universe is the start of everything. Before the universe there was nothing, so the underlying concept is that it must be built from nothing. This requires real 'lateral thinking'. There was a story many years ago of a construction project in which they needed to build higher than the tallest crane. They thought, how can we do this? Someone said, 'We need sky hooks!' From this slightly mad comment came the idea of using a helicopter to bring up the building materials.

The creator of the universe might have said, 'We need to build our universe from nothing, so let's have matter and antimatter. These can cancel each other out to be nothing'. With this idea a universe could be born.

The next simple idea is to make everything from atoms. We know that atoms are extremely small, in the way that 'extremely small' does not do justice to how small they are. We also know that the universe and the things in it are extremely big, with an extreme amount of bigness. Our Sun is one million times the size of Earth, but it is but a speck when compared with stars that can fit 3000 trillion Earths inside. There are an estimated 200 billion galaxies, containing on average 100 billion stars each, plus planets and moons. It is certainly imaginative to build everything from tiny atoms, but this approach gives the simplest design.

It means that there are two stages in building a universe. Stage one is to create the building materials, namely the atoms, oh, and just do this from nothing. Stage two is to use them to build your universe. Simple really.

We can have a go at understanding how to create a universe and planet life, but who or what did the creation is beyond our understanding. How about *why*? No, drawing a blank on that one too. However, if you knew how to build a universe but never got the opportunity, that would be so frustrating.

For the purposes of our story, we will imagine the ridiculous idea that the Creator has teams working in the background on the creation tasks. There is the Universal Building department, who build universes, and the Planet Life department who create life on planets. They can help explain some of the important points.

SECTION 1 - UNIVERSE

How to Make a Universe

Let us start by looking at how the universe was created. In Section 2 we will turn our attention to planet life.

Obviously for us planet-dwellers, universes are outside of our experience, and push the limit of what we can really understand. However, we should not be put off. We can take it gently and build our understanding.

Let's drop in on Universal Building to see if they can help us understand how the universe started.

1.ONE

Universe from Nothing

They are holding one of those inter-departmental meetings, the sort where one department explains to the other what they do. It is the turn of Universal Building to talk to Planet Life.

'Why did you decide to build a universe?' someone from Planet Life is asking.

'It wasn't our idea, it was the Creator's,' comes the reply. 'And we asked the Creator the same question. 'Why do you want to build a universe Creator?' And the Creator said, 'Because I know exactly how to do it, so if I didn't build one, I would feel massively unfulfilled.' And then the Creator continued with a smile, 'Anyway, I am absolutely brilliant at designing and building things, so it will be fun!'

Not understanding the true scale of what was being proposed, someone had asked innocently, 'Will the universe be quite small, tabletop sized maybe?'

And with infinite patience the Creator had calmly replied, 'No it will not. It will be absolutely huge with trillions of stars, planets, moons, and other mysterious wonderous things, because the top requirements for a universe are being big and full of wonder and mystery.'

And everyone had had the same thought, how is the Creator going to get all the quadrillions of tons of material to build such a huge mysterious wonderous universe.

'I will try and explain how the Creator did it,' says the presenter, ushering his audience to come to a door which opens onto a small room with nothing in it. 'What do you see?' he asks.

'An empty room,' says one person. 'A room with nothing in it,' says another.

'Precisely,' says the presenter, now taking a chair and putting it in the room. 'What do you see now?' he asks.

'A chair,' says one person. 'A room with a chair in it' says another.

'Exactly,' says the presenter and continues. 'Let's think about this. The reason I did that, was to make us think about what we really mean by *nothing* or *empty'*. 'Nothing,' he continues, 'means the absence of something. In this case the absence of the chair. You cannot have the concept of nothing without the concept of something. Do you agree?'

'Well, sort of I guess,' says one of the Planet Life team, 'but so what, why do we care about that?'

'Well, prepare to be amazed!' he says, ushering them back to their seats. He holds up two identical jars, each of which was half-full, and both had labels on.

'What do you see?' he asks.

And someone at the front peered forward to read the labels. 'I see two identical half-full jars, one of which has a label saying *Matter* and another with a label saying *Antimatter*.

'OK, so if I pour the contents of one jar into the other, what will happen?'

'Obviously, you will have one empty jar and one full one,' comes the reply.

'OK, let's try it'. He starts pouring the antimatter into the matter jar being extremely careful not to spill any on his foot, and they were amazed. Instead of going up, the level in the matter jar was going down, until the presenter was finally left holding two empty jars.

'Isn't that amazing!' exclaims the presenter triumphantly. And everyone sits in stunned silence, not sure what they had seen, until someone asks, 'Why did that happen? Where did all that stuff go?'

The presenter replies, 'We had something in each jar. And now we have an absence of something. We have nothing! Two empty jars.'

Passing the *Matter* jar around, they notice how freezing cold it has become.

And he continues, 'This is the nature of matter and antimatter. They annihilate each other. Weird isn't it? You probably wouldn't believe it unless you had seen it for yourselves. Add antimatter to matter and there is no longer any matter or antimatter. There is just nothing; the absence of something.'

Pausing for a moment to take it all in, someone then asks, trying to get their head around it, 'This nothing which is in the jar, where the matter and antimatter annihilated each other, is it different to the nothing that is in the other jar?'

'No,' says the presenter. 'Nothing is just nothing! There aren't different types of nothing.

'And out in space we have an infinite amount of nothing. Cold, dark, empty space going on forever in all directions.'

'I get it,' says a quiet, smart one sitting near the back who had been silently taking it all in. 'You can do the reverse of what you have just done. Oh, that is clever!'

Most of the others were looking on blankly, so she continues, 'You can take nothing and turn it back into matter and antimatter'.

'Well, to be accurate, turn it into matter and antimatter for the first time,' corrects the presenter.

'And,' she continues getting more excited as the possibilities dawn on her. 'There is an infinite supply of nothing in space. All the nothing that you need to make the universe as big as you want. And you can make as many universes as you want! That is clever! That was all the Creator's idea, right?'

'Yupp,' says the presenter. 'We found it amazing when the Creator explained it to us.'

After a suitable pause to let it all sink in, the presenter continues, 'There is just one more thing that we need to think about, and that is, how big is nothing? How much space does nothing take up?

'After I poured the anti-matter into the matter, how much space did the resulting mixture take up in the jar?'

'It didn't take up any space. The jars were both empty.'

'Correct,' answers the presenter. 'If I had used jars that were twice the size with twice as much matter and antimatter, then how much room would the resulting mixture have taken up?'

'It would have been the same. It wouldn't have taken up any space.'

'Correct,' answers the presenter again.

'If I had taken matter the size of a star and mixed it with equal amounts of antimatter, what about that?'

'Same again,' answers the team in a slightly incredulous way.

'Are you saying then,' asks the smart one, 'that even though space is infinite, nothing takes up no space at all?'

'Yes, right again. I know it is a bit hard to get your head around. We could take a planet and a star and then another star and they would all fit into the same amount of space, which is no space at all.'

'So,' she continues. 'When we go the other way. When we create stuff from nothing, however big the stuff we create is, it always starts off the same size, which is no size at all.'

'Correct! A bit weird isn't it?'

They sat in silence for a while, and then the presenter announces, 'Let's take a coffee break!'

Good idea, they all think. We need a break after that.

Oh, they have laid on doughnuts. That is thoughtful of them, thinks one of the Planet Life team. Meetings always go better when the host takes the trouble to get sticky buns or doughnuts. When it is our turn, I will do my famous chocolate brownies in the shape of some of the creatures we are creating for our latest planet project.

1.TWO

The Beginning

We can understand relatively easily that we cannot have the concept of something without also having the concept of nothing. They are opposites, like light and dark. However, it is not just a concept. It is literally impossible to have something without having nothing. Let us think about this.

When we create something on our planet, we always start from something else. I want to make a wooden table, so I start with pieces of wood. The plumber is installing the central heating and comes with pipe and fittings. This is how it works in our world; we change one thing into another thing.

When you create a universe, it is different; you are building something for the first time rather than changing one thing into another. It is the ultimate 'greenfield site', or rather 'dark, empty space site'.

We can picture dark, empty space where there is nothing, no heat, it is absolute zero, no light, just darkness, and nothingness that goes on for ever in all directions. Physically the universe must exist in this nothingness. The nothingness that goes on for ever in all directions, was there first. The somethingness of the universe was created inside it.

Philosophically too, and this is harder to grasp, to have something, there must be nothing. This is worth thinking about because it is a bit strange, and it takes time to really understand. When people propose that maybe there were some sub-atomic particles hanging around before the Big Bang, that comes from thinking in our familiar terms where we always start with something to make something else. It misses the point that the universe is the start of everything and therefore it must be created from nothing. Something cannot exist without there being nothing first.

When we create something there are always constraints. You want an extension on your house, there are planning rules, financial constraints, the type of materials to use, and so on, but, with the universe, there are no constraints. This is for a simple reason. When you create a universe from nothing, nothing exists, you must design everything. Obviously, once the universe has been created, everything works according to the way it was designed, but at the beginning, when there was nothing, everything had to be worked out.

It is an amazing opportunity for our creator to be able to start from nothing and have no constraints on the design. Look how big the universe is, there are no constraints on its size. The downside is that the creator of the universe must create everything, must work out how it all works, how it all fits together, no modules, or widgets, exist to build the universe from.

We say 'no constraints' but there is one constraint. The universe must be built from nothing. How do you do that? It is the brilliantly simple idea of having matter and antimatter, which, in equal quantities, cancel each other out. Einstein's $E=mc^2$ equation shows how this works mathematically, because if there is *m* meaning normal matter and -*m* meaning antimatter in equal quantities, then *m-m=0*. There is no matter and no energy either, which is the definition of nothing.

We can show mathematically that matter and antimatter will cancel each other out. Weird as this seems, scientists working with the Large Hadron Collider in CERN, have been able to demonstrate for real that matter and antimatter do indeed cancel each other. Zap them together and afterwards there is no matter and no antimatter; they have gone. As a side point, the antimatter used in the experiments comes from radioactive decay of matter, so there is antimatter inside some normal matter.

Let us think about this. We can show mathematically that matter and antimatter cancel each other, but when there was nothing there was also no mathematics. Maths had to be created too. Indeed, the universe has been traced back to 10^{-49} second after its creation but not back to zero seconds, because at that point there was nothing including no mathematics or laws of physics. Maths describes things like size, speed, volume, distance, forces, and time; all the fundamental properties that are needed when creating a universe. To prove that matter and antimatter cancel each other, we use mathematics, but mathematics also had to be created.

We have understood maths for ever. Imagine a family of cavemen eating a meal, and the young daughter uses some grunting language to say that her brother has been given two bones to eat and she only has one. 'Yes,' grunts back the mother. 'But your bone is twice as big as your brother's.'

We understand maths, but do we stop to consider how it is something we discovered? Maths existed as part of the universe and we discovered how it worked, we did not just invent it ourselves. Maths works in a unified way; the results of a calculation must always be the same whatever calculation method is used.

But what if it wasn't like this? What if you could calculate in different ways and get different results?

Mathematics

A woman gets to the front of the queue in the bank and goes to talk to the next available cashier.

'Can I help you?' asks the cashier with a polite smile.

'I certainly hope so,' says the woman in a slightly agitated state, and she holds up a copy of her bank statement so that the cashier can read it through the glass.

'I recently opened this account and deposited £10,000. Over the month I have made some withdrawals. You can add them up to see that they come to around £1,000.'

Trying to keep a business-like demeanour, she continues, 'My balance should be around £9,000, but look at the balance, it is showing £251.26. There has been a terrible mistake and I want it rectified.'

'Oh, I see,'. says the cashier putting on a serious, concerned face. 'Let me look into it.' She taps away on the computer and then walks off to the printer returning with another bank statement print out, which she pushes under the glass screen to the customer. 'No, it all seems to be completely correct,' she announces.

'What!' screams the woman incredulously.

'Yes,' continues the cashier calmly. 'The difference between the amount you were expecting and the amount we have calculated is down to maths. Maths, as you probably know, varies hugely depending on which method of calculation you use. The bank uses a method that makes your balance as low as possible, and our profits as big as possible.

'To explain, I have printed an extra column on your statement headed *BN* which stands for Bankers Number, which is 561.32.

The woman stares at the new statement with the extra column, desperately trying to understand in a state of shock, and the cashier continues.

'As you can see, we multiply all the transactions by the Bankers Number, add them up as you just did, subtract from £10,000, and then divide by the Bankers number again. Calculated in this way, we get a total of £251.26. So, I am pleased to confirm that it is all correct.

'Is there anything else I can help you with today? Maybe you would like to make a cash withdrawal?' she asks with a smile. 'You should be able to withdraw £20 without going overdrawn'.

'What!' continues the woman ignoring the last question and looking at the cashier again. 'Do you always calculate in this strange way?'

'Well, yes on customer accounts we always use this method.

'I believe that this method has been used for years by multi-national tech companies to work out how much tax they owe. It can also be used in reverse, for example to calculate the bonuses of senior executives. Divide by the banker's number first and then multiply afterwards to make the number as big as possible.'

And continuing, 'It is all clearly set out in the Bank's Terms and Conditions, so there is really nothing more I can do about it I am afraid.'

'But I read all the Terms and Conditions before opening the account. I am always very thorough like that, and I didn't see it anywhere.'

'I can assure you that it is there, madam. Log in to internet banking, find the T&C's page and then scroll all the way down. Here you will find the *Additional Terms and Conditions* link where the calculation method is clearly explained.'

One of the cashier's colleague chips in, 'Yes that is right, but on some computers, you may find that you cannot see the link even though you have scrolled to the bottom.'

'How do you mean?' asks the woman in a puzzled voice.

'On some computers you cannot see it unless you also use the down arrow key.

'Actually,' she continues. 'When the developers did the website, they used an obscure screen aspect ratio with a never-used screen resolution, and they could see the link on their screens, but unfortunately it seems that nobody else can see it on theirs.'

'But that sounds like a deliberate attempt to hide what you do from the customer! This is outrageous. I will log into my internet banking, scroll to the bottom and then use the down arrow key, and check out what you have told me.'

'Not just the down arrow key madam,' adds the colleague absent-mindedly.

'I beg your pardon?'

'You have to use one of the function keys at the same time. I cannot remember which one, maybe F4 or F5 or F9 possibly'.

'Well, alright. There aren't many function keys, so I will try it,' she says tersely.

'And the other key somewhere on the bottom left of the keyboard,' adds the cashier in a disinterested manner.

'Yes, you have to press them all simultaneously a bit like *control, alt, delete* but much more sensitive. You have to push them all at exactly the same time or it won't work'.

'And,' continues her colleague 'you need to have rather large hands to reach all the keys simultaneously. I believe they tested it by using a concert pianist who had unusually large hands. Other people generally cannot manage to reach all the buttons at the same time.'

'On the training course, I used my big toe on the down arrow,' says her colleague.

'But you're a freak; all that gymnastics when you were younger,' says the first cashier. 'I used my nose for the function key and my left and right hands for the other keys. I couldn't see what I was doing, or see the screen, but somehow I got it to work eventually.'

At this point they all spontaneously duck their heads down onto their keyboards doing an impression of using their noses on the keyboards, which they all find hilariously funny, but the woman doesn't appear amused at all.

The woman continues as calmly as she can manage, 'So, if I somehow manage to find this hidden link, what happens next?'

'Oh,' they say almost together. 'There is a screen, and a message says: *This link is totally unsafe, are you sure you want to continue?* It says there is a danger of phishing… ransomware… junk emails… scams… viruses… trojans…,' trying to remember what else was in the list.

'Some people find that a bit off-putting when they are logged into their internet banking, but just ignore the message and press *Yes* to continue.

'That takes you to another screen that says: *Are you really sure that you are not about to make the worst mistake in your whole life by continuing?* That one has a rather nice picture taken from an Elizabethan woodblock print showing people being tortured and disembowelled in the Tower of London.

'Just press Yes on that one as well to continue, and you get to the Further Terms and Conditions page where everything is clearly explained.

'Out of interest madam, do you also have a mortgage with us?'

'Luckily no! Why?' says the woman tersely.

'It was just that we had a message from Head Office saying that they are now using a new mathematical algorithm to calculate the mortgage repayments. Some people are noticing that however much they pay off, their balance just keeps going up, and then their house gets repossessed'.

Once there was mathematics there could be things that need maths such as time, distance, and forces. Consider this too, there must be a complete design in which everything fits together in a coordinated way.

We intrinsically understand that the universe, and everything in it, is part of one complete design in which everything has been worked out, and everything fits together.

Maybe Einstein used maths to discover $E=mc^2$, rather than just work it out. Maybe this is a fundamental equation used in the design of the universe. It was Einstein himself who observed that the best design is the simplest one that works. We have a feeling that something must be right when the answer is simple.

In designing a universe every single detail must be worked out. It is much easier when every piece of the puzzle is simple. This way the layers of complexity can be built from many simple building blocks, and this is possible because there are no constraints, everything can be done in the best and simplest way.

Returning to creating something out of nothing, and the beautifully simple concept of matter and antimatter. It would be tempting to think that there must be only one type of matter/antimatter pairing – the *God Particle* as it is known. Maybe there is because this might be the simplest design. However, there does not have to be, because this is not a constraint. It might be that the simplest method is to have a mix of sub-atomic particles coming out of the nothingness in exactly the right proportions. The thing is, and this is difficult to grasp, the matter/antimatter pairs do not actually exist in the nothingness, because there is nothing. Instead, they are created for the first time from nothing and in the end, they will annihilate each other and go back to nothing – as observed for real in CERN.

1.THREE

Atoms

There is a meeting going on before any universes have been created. Written on a whiteboard behind the Creator's head is 'How big should we make an atom?' the subject of the meeting.

One of the universal builders, who has been studying the plans for the first universe, kicks off, 'It says here that we have to build quite a lot of stars that are 3000 trillion times the volume of Planet Earth. That means that if we made an atom the size of Earth, we would need 3000 trillion atoms to build those biggest stars.'

His mate then chips in, '3000 trillion atoms, that is a lot of atoms, do you think we could really do that?'

The senior builder does that thing that all builders do, and sucks in through his teeth, nods his head gently and adopts a knowledgeable expression, 'Well, it will be hard, that is a lot of atoms, but I think we could do it.'

'What about the moon?' pipes up another team member. And when the others look puzzled, he continues, 'According to the plans, the Earth has a moon. Looking at the size, 50 moons will fit into the Earth.'

'We will have to make an atom the size of the moon then. 50 atoms for the Earth, and now 150,000 trillion atoms for the biggest stars.'

More sucking of teeth and 'tut tutting'.

'I need to make an ant!' announces a woman from Planet Life department, affectionately known as Mother Nature. 'And before you get any ideas,' putting up a picture of an ant on the old-style overhead projector, 'I am not talking about a round ant with a smiley face looking remarkably like one single atom, I am talking about an ant with lots of sections and legs and other stuff.'

This is met with stunned silence.

Eventually, a hopeful voice says, 'Could we drop the ants and say that the smallest creature is an elephant?'

Mother Nature's voice comes back, 'I suppose we are talking about a big round ball-like elephant, looking like it is made from one single atom?'

'Oh, that is a good, idea... why didn't I think of that?' returns the first voice.

At this point, the Creator, who had remained silent, listening to the conversation, steps in.

'We cannot have big atoms for making stars, planets and moons, and small atoms for making life. It doesn't work like that. There must be just one size of atom for everything.'

'Why is that?' asks one of the universal builders hopefully.

'Because,' continues the Creator. 'All my designs are brilliantly simple,' smiling in a pleased way. 'We will have one type of atom for everything.' Yes, there will be different atoms like hydrogen, helium, lithium, and so on, but they will all be made in the same way. One type of atom for all the matter that we will create in the whole universe.'

'Now,' the Creator goes on. 'This idea of making the smallest creature the size of an elephant. That I am afraid is not going to make any difference.'

'Why is that?' enquires a universal builder.

'Because', says the Creator, 'Life is another one of my brilliantly simple designs,' smiling at the thought of how absolutely brilliant this one is too.

'Uh oh here we go,' seems to be muttered under various breaths.

The Creator continues, 'All life is made in exactly the same way. Isn't that clever?' now warming to the task of explaining the design, as all designers do when explaining the designs they are proud of. 'Every living organism will be made of cells. The simplest organisms will be just one cell. When they reproduce, which is what living things do, they will create a copy of themselves and then there will be two organisms that each have one cell. But – and here is the especially clever bit – we can get more complex multi-cell organisms through the same mechanism. Instead of splitting and becoming another single-celled organism, the cell will split and become another cell within the same living creature. We can build all sorts of things like trees and grass and ants and elephants and humans all with the same design. Brilliant!

'And to make it alive each cell contains genetic code that has all the instructions on how to build, how to operate and how to maintain that cell.' Every cell has this code because that is the design, either single cell or multiple cell organisms, the pattern is always the same'.

'What is genetic code?' asks one of the builders.

'Atoms,' replies the Creator. 'Atoms are joined together in a molecule complex enough to hold all the instructions, and small enough to fit inside a single cell.'

'How many atoms will we need in a genetic code molecule?' asks a builder with some trepidation.

The Creator replies, 'About 200 billion should do it.'

And getting up to leave the room, the Creator says, 'Now that this is all clear, I will leave you all to finalise the details.'

'What about my stars that are 3000 trillion times the volume of planet Earth?' mutters a despairing builder under his breath.

'We will just need lots of atoms when we create the universe. Simple really.' Says the Creator.

I used to think about how there are cosmologists, astrophysicists, astronomers, and others who are observing, and trying to understand, the universe, the stars, the planets, and all the huge things. Meanwhile there are quantum physicists who are looking at the tiniest things: atoms and sub-atomic particles. Could, for example, the Moon orbiting the Earth shed light on how an electron orbits a nucleus?

But it is simpler than that; everything from the biggest stars to the tiniest cell in a bacterium is made from atoms.

Imagine the pharaoh and his architect discussing how to build the Great Pyramid at Cheops. The architect says, 'As the pyramid is in the desert, I think we should build it from individual grains of sand. We can make glue from camel bones and the builders can stick each grain of sand together trying not to get sand stuck on their fingers.' The pharaoh says, 'I don't think that is a good idea at all, when we build something big, we should use big things like blocks of stone.'

Here is the extraordinary thing, there are billions of atoms in a single grain of sand. The Great Pyramid is 139 meters high, whereas a star made of atoms can be 175 million miles in diameter.

It is not immediately obvious that you would build things so big from things which are so small.

That is definitely thinking outside of the box. But, if you can do it, then why not? It goes back to the best design being the simplest one that works. It is simple to build everything in the universe from just one thing, the complexity comes from the scale.

Those who have done any sub-aqua will know that at 10m the water pressure is twice atmospheric pressure, at 20m it is three times. A diver must equalise the pressure in their ears to avoid burst ear drums. What is the pressure like in the middle of the Earth, or in the middle of a star 175 million miles across?

To build a tiny atom that can take loads like this means that the design must be perfect. And this is the point, the design is perfect. Imagine a scientist giving a paper at a conference, 'I have understood how an atom is designed, and quite frankly it is a bit rubbish, I wouldn't do like that. Maybe this was an early version of a universe and there have been some improvements made in later universes.' It is not going to happen is it.

Forces

Just four fundamental forces govern how all the objects and particles in the universe interact with each other (although a fifth force may have been discovered): gravity, electromagnetism, the strong force and the weak force.

This simplicity comes from making everything in the universe from atoms. Gravity keeps us from floating off the planet, because we and the planet are both made of atoms.

1.FOUR

Creating the Atom

We know that the first stage in building a universe is to create the materials to build it from; namely the atoms. Universal building materials are not just hanging around in space. Indeed, we have discussed how it is the one constraint on the design that it must be built from nothing.

Scientists talk about epochs in the creation of the universe. These define what happened during a particular period. However, these can also be viewed as stages in a giant industrial process for the manufacture of atoms.

When a pharmaceutical company create a drug, they call it 'creating the molecule', because this is what a drug is, a molecule that has been designed to be put inside the body to fight a disease. It is a process of building a molecule from atoms.

How do they do this? They start with different raw materials in the desired quantities. They can change the temperature, vary the pressure, and vary the time spent in each stage of the process. The process will have been worked out in a way that is repeatable so long as everything is done the same way each time. An American company could use manufacturing facilities in Mexico and in India, for example. These facilities follow the same recipe.

The first stage in the creation of the universe is, 'creating the atom', and we can see that it is like the pharmaceutical 'creating the molecule' process. The quantities and proportions of raw materials could be varied, these are the fundamental matter and antimatter subatomic particles. The volume of these, together with the size of the universe, determine the heat and pressure. The final variable is the time spent at each stage before the universe changes size, both to give room for the volume of the ever-expanding particles, as well as to reduce the pressure and temperature.

1.FIVE

Big Bang in a Ball

Imagine you were in a spacecraft and you travelled to a location far away from the gravitational forces of any stars, planets, or moons. Here, you let off a firework. What would you expect? Wouldn't you expect that the pieces of firework would fire off in all directions and, without any gravity to divert their path, or any atmosphere to slow them down, they would keep on going for ever?

The universe starts with the Big Bang, but a big bang just in empty space would do the same as the firework. All the particles would bang off in all directions and continue for ever. There would be no chance for them to join with other particles and thereby build into atoms.

The Big Bang does not happen just in empty space, rather it is set off inside the universe. The universe is an entity; we can imagine that it is a sphere because that is the most efficient shape. Amazingly, at this time, the universe was the size of a football. It was the most incredible pressure vessel because it contained all the matter and all the energy for the life of the universe.

The earliest stage can be traced back to 10^{-49} seconds after the Big Bang, so, how long does it take for all the fundamental subatomic particles to come out of the nothingness? Logically, it takes no time at all, because there is no concept of time before the universe is created.

At this point, the particles must be infinitesimally small and densely packed, because the quantity of particles required is so astronomically huge and the size of the universe is ridiculously tiny. Bear in mind that its current size is estimated to be 93 billion light years across.

The earliest epochs, or stages in the atom manufacturing process, are called things like the Quark, Hadron and Lepton epochs which are the names given to sub-atomic particles, implying a sequence in the manufacturing process.

Surprisingly, for a universe that is currently nearly 14 billion years old, the stage of having basic hydrogen nuclei built from fundamental particles is believed to be reached after just 20 minutes.

The universe started as a 'singularity' because it is made from nothing and nothing takes up no space at all. It then became the size of a football, before growing to the size of a spherical city centre office block and, with each stage completed, it expanded again. The pressure vessel changed size, both to provide the ridiculously high pressures and temperatures needed for the first stage of atom manufacturing, and to accommodate the size of the particles as they expanded through the manufacturing process.

1.SIX

The Elements

Let us get back onto somewhat more familiar ground. Atoms mean the elements in the period table – hydrogen, helium, lithium, beryllium, boron, carbon, and so on. We have known, for what seems like forever, that an atom has a nucleus, where protons and neutrons are found, with electrons orbiting it.

An Atom

How many different elements – meaning different types of atom – are there? Is it several billion, as the universe deals in such large quantities? No, there are just 94 naturally occurring elements on Earth.

As we go up the periodic table, it simply means adding one more neutron, proton, and electron. Just three things; there are not different types of electrons, neutrons, or protons. It is quite remarkable that by adding one more of each to an atom such different elements are created, all the elements that we are so familiar with in our daily lives – oxygen, carbon, iron, copper, aluminium, and so on. Gases, liquids, and solids.

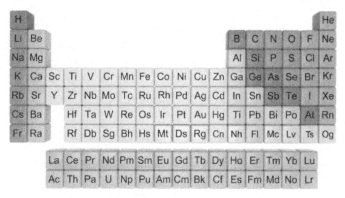

Periodic Table

The impressionist painter, Monet, used just nine paints. These were lead white, chrome yellow, cadmium yellow, viridian green, emerald green, French ultramarine, cobalt blue, madder red, and vermilion. Our creator uses just 94 naturally occurring elements to 'paint' everything in the universe. Like a painter, the basic palette of elements can be extended by mixing them together with other elements. Do we ever say that there are not enough? 'Sorry I cannot redecorate the bathroom because I don't have the right atoms?' No, we have all the atoms that we need.

And what about the quantities of the atoms? We have large numbers of the ones that we need most of like iron and copper and aluminium that are used for construction, and fewer of the ones we consider precious and shiny like gold and silver, which keeps these ones valuable.

Diamond Ring

A young woman dressed in the finest quality clothes with beautiful hair and make-up is sitting elegantly in an expensive London restaurant. She is about to make use of the Chancellor's 'Eat out to Help out' deal which will save her £10 on her hugely expensive lunch. The waitress approaches to take her order.

'Look at my amazing diamond ring,' says the young woman, proffering an elegant, manicured hand.

'That's nice, did you get it at Lidl,' comes the somewhat unexpected reply from the waitress.

'Did I get it in Lidl!!?' screams the young woman with 1996 Dom Perignon Rose Gold champagne spraying out of her mouth. 'What absolute cheek! This, I'll have you know is called the '3 in 8 Billion' diamond. It means that there are nearly 8 billion people on the whole planet and only me and my two bestest – and richest – friends have got one.'

'I wasn't meaning to be cheeky,' says the waitress. 'Maybe you haven't heard that they found a gigantic seam of diamonds underneath Lidl's car park in Neckarsulm, Germany. There are now so many diamonds that anyone can have one. They were selling diamond rings in the *Middle of Lidl* for £15 each, and chef bought each of us waitresses one. Come on girls lets show her our rings.'

The other two waitresses approach, and all three together hold out a ring-fingered hand, 'Dun nah look at our lovely diamond rings!'

And peering closer to have a better look, the waitresses say, 'Ours seem to be a bit bigger than hers. And look, the purity of our diamonds is slightly better. But otherwise, they are quite similar.'

At this point the young woman starts screaming. 'I hate you all. I hate this restaurant. I never want to come here ever again. I know... I will get Daddy, to shut you down.'

And the waitresses looked perplexed, 'Strange? Why is she so upset and angry? I thought she would be so pleased to know that we all had one. All girls together in the diamond ring club!'

How is this? How come there are exactly the atoms that our planet needs and in exactly the right quantities? It is all part of the plan.

I am reminded of a lodge in Nepal, the last stop for people trekking up to Annapurna Base Camp. Although they had limited ingredients, they provided such amazing food to the hungry trekkers that it became famous on the backpacker circuit. You would look at the menu, but then the Captain would come and say, 'May I recommend the Captain's potato soup, or the Captain's Pizza, or the Captain's Apple Pie?' And you would just say yes to whatever he recommended. From flour, eggs, potatoes, and the occasional apple, such an array of delights could be conjured up.

1.SEVEN

Atom Expansion

Imagine you are on an extremely long beach. You are carrying a basketball. The basketball represents the size of the nucleus of a hydrogen atom. You put down your basketball and you walk for 2 miles (3.2km) down the beach, which, for most people, is a 40-minute walk. At this point you would be whacked in the head by an exceptionally large electron orbiting your basketball nucleus.

Two miles represents the radius of the atom. If you had a lot of basketballs and you laid them down next to each other in a line for that two-mile distance, you would need 133,000 of them. To fill in a disk on the sand with a two-mile radius would require 33 million basketballs. To make a complete sphere with a two-mile radius would require more than 2 trillion basketballs.

An electron orbiting a nucleus pumps up the volume of the smallest atom, hydrogen, by 2 trillion times. That is an incredible expansion, especially when we consider how many atoms there are. As the process of making atoms goes on, the universe expands to fit the increased volume of everything in it.

Planning Department
A man is in a meeting with his local Planning Department. The planning officer is saying, 'I see that you currently live in a shed, but you want to expand into a house.' The man concurs and the planning officer continues. 'It says here that you want the house to be 4 miles wide by 4 miles deep and 4 miles high. Surely that must be a mistake.' 'No, that is absolutely correct,' says the man.

'Do you have a large family then?' asks the planning officer. 'No, it is just me and the dog,' replies the man. 'Then do you have a lot of furniture?' asks the planning officer. 'Not really, just a sofa and a cabinet for the TV,' comes the reply. The man continues, 'I just like a lot of space. I will sit on the sofa watching TV two miles up and two miles from the outside walls.'

'You do realise that to bring the shopping up the stairs will involve climbing more than the height of the UK's highest mountain – Ben Nevis – from sea level?'

'And,' says his colleague. 'Spare a thought for the window cleaner trying to clean the windows with those telescopic poles.' Would it not be better to build something a bit smaller?'

'Actually,' replies the man, 'This will be my smallest house. I am planning some others that are three times the size.'

Let us think about this, our Moon orbits the Earth. That does not mean that the diameter of the Earth got bigger. It is just a planet of a certain size with a moon orbiting around it. Something else must be going on in the atom to make it grow in this way.

It was discovered some 80 years ago that electrons can be in more than one place at once. Indeed, there is only a probability that an electron is at a given location at any time. This phenomenon must be important for the expansion of the atom.

Do you remember the old-style televisions with cathode ray tubes? The back of the screen was coated with a phosphorescent coating. A beam of electrons was fired at the top corner of the screen causing the coating to light up. The beam scanned across and then moved down to the next line and scanned back along that line until the whole screen had been beamed. This was done sufficiently fast that the coating in one place had only just started to fade when it was hit with the beam again. This gave the impression of the whole screen being lit up even though there was only one beam of electrons.

Could something similar be happening in the atom? The electron is in an orbit, but it is not in only one place, it is everywhere at the same time to create a complete sphere at the larger radius. This would be a brilliantly inventive way of expanding the volume.

Brick

When we look around us, we see everything that we are familiar with, everything looks normal. It is only because we have been inquisitive and have looked inside atoms that we have an idea of what is going on, which seems to us not to be normal at all. But that is inevitable is it not? To build a universe from nothing does not feel normal, it feels distinctly strange. It is clever to go from distinctly strange to normal in a sequence of steps.

A builder walks into a small, private builders' merchant. He says, 'Hello, I need something to build the wall of a house.' And he continues, 'It needs to be solid and strong to support the weight of the wall and the weight of the rest of the house - you know - the roof, the floors, and stuff inside the house like beds, cupboards, people and maybe a large dog.'

'I have just what you need,' replies the salesman holding up a brick and passing it to the builder: a brick!'

The builder weighs it up in his hands and feels it. 'Looks good,' he says. 'It seems solid, heavy and strong.'

'Seems,' replies the salesman.

'Pardon?' says the builder, raising an enquiring eyebrow.

'Yes. It seems solid but actually it is 99.9999…'

'OK. I get the picture,' interrupts the builder slightly impatiently.

'…9999%,' finishes the salesman, 'empty space.'

'Riggggt,' says the builder slowly. 'So, the other 0.0…….00000001% or whatever must be solid and heavy.'

'Not really,' replies the salesman with a smile. 'Sub-atomic particles are not famous for being heavy, that is more down to the invisible force of gravity. And another thing, the sub-atomic particles are whizzing around inside this brick at extremely high speed,' he says taking back the brick, holding it to his ear and shaking it to try and hear something.

'Look,' says the builder, 'I am a practical guy. I can see what I can see. That brick is heavy and solid and is just the job for what I want even though you have done your best to put me off. I am not put off because I trust my years of building experience. However, if the Building Regs people get hold of what you have just said, they may use it an excuse to say that anything made with bricks is unsafe just to create unnecessary jobs for the construction industry. So, I am not going to take the risk.'

And with that he walks out of the shop, leaving a slightly surprised salesman thinking, I have been looking after the shop while my mate is away for two weeks holiday, and I haven't sold a thing.

1.EIGHT

Light

I once owned something that looked like a traditional lightbulb containing a vacuum. Inside, balanced on a needle, to give low friction, was an assembly of 4 vanes, painted black on one side and white on the other. When put in the sun, the vanes would rotate, sometimes going like the clappers. It is called a Radiometer; you may have one or know what I am talking about.

A Radiometer

Why do the vanes rotate when the light is absorbed by the black side of the vanes? The simple explanation is momentum; momentum is mass times velocity (mv). It demonstrates that light has mass. It has tiny mass and the speed of light, which is enough to push the vanes around.

This demonstrates that light is photons, which is not a problem for light, except that we always thought that light was waves, making it a problem for us. Once we got over that, we had the issue that light was observed to pass through two slits at the same time, meaning – to us – that it could not be a particle because a particle could not be in two places at the same time. Except that was not a problem for light either, so we had to get over that one as well. Light was always cool about being waves and particles, and in more than one place at once, just like electrons. It did not bother light at all.

Einstein's $E=mc^2$ – if indeed it is Einstein's – 'sheds some light' on this because it does not just say that energy is mass x the speed of light squared, it also says that energy *is* mass and mass *is* energy.

$$E = Mc^2$$

Albert's Dream
We say 'if indeed it is Einstein's' because it could be an equation used in universe building that Einstein discovered.

We can ponder whether any of us are really clever enough to work these things out ourselves. I believe Einstein had the experience that many of us have of a solution 'coming to us in the night', some level of clarity, a Eureka moment.

Let us imagine that an angel was sent to Einstein to help with working out $E=mc^2$, but unfortunately first time around she accidentally went to the wrong Albert.

It was early morning and Albert Entwistle, the Yorkshire street sweeper, was with his colleague Fred. They were both leaning on their brooms at the top of a steep cobbled street with houses crowding one side and countryside beyond, because that is what Yorkshire looked like in the early 20th century.

'I had a weird dream last night, Fred,' announced Albert, keen to share it with someone.

'Oh yes?' said Fred, surprised because Albert had never, to his knowledge, had dreams before.

'Yes,' continued Albert. 'This voice came to me and said something like, *Dee antvort duss zee zuken…*'

'Bert! You are a queer one. The first time you ever tell me about a dream, and you spout off a load of foreign sounding gobbledygook.'

'I am just repeating what I remember from the dream, Fred. It was weird! But I think I must have subconsciously said something like: *You what luv?* because the voice sounded a bit surprised and switched to English: *The answer you are looking for… is $E=mc^2$*'.

Fred repeated in a puzzled way, '*The answer you are looking for is $E=mc^2$*. What answer are you looking for Bert? What can that mean?'

'I dunno. No idea. I am not aware that I am looking for any answer to anything'.

And Bert continued, 'I think I must've seemed a bit surprised in my sleep too. It was like the voice expected me to understand, but it continued: *Ja, I mean yes, that really long, complex equation on page 24 of your workings, with lots of Greek letters, integration symbols and curly brackets, is wrong. The answer is in fact very simple, and with this knowledge your name will forever be associated with being a genius'.*

At this point, Fred fell about in uncontrollable laughter. 'Your name forever associated with being a genius!!! I am really sorry Bert, but you are known for being as daft as these brushes. When folks round here do something stupid, they call it: *Doing a Bert!*'

'Oi Fred, that is not fair. I resemble that remark!'

Is this not a fundamental principle? These particles that come out of the nothing (as also described by the equation) are a mixture of energy and mass; they are both things together. When the Big Bang went off, it did not release all the mass for the universe and all the energy for the universe separately, it released particles that contain all the mass and all the energy together. Another simple concept.

We can get an insight into this when we boil water in a saucepan. The energy from the heat agitates the water molecules releasing their bonds and turning them into steam.

Where do we get our energy from? From burning wood, coal, and gas, from uranium, from the power of water and wind. This is all energy from matter. The exception is solar, which is both energy and matter. It is solar that powers life on Earth

1.NINE

Making Hydrogen

We are now ready to put the pieces together, going from making the most basic atoms through to using them to create galaxies and stars. Hopefully, you are still up for thinking about these complex things.

We will be talking principles here. Clearly there is a great deal going on to create everything needed for a universe, so our discussion is a simplification. Nevertheless, our creator does simple designs based on simple principles so there is insight from keeping our thoughts simple. Do note that the following is not necessarily right, this is the universe after all, but perhaps it can help us to think about the universe and what it is all about.

The first stage is to make the simplest atoms.

Lab at the Beginning of the Universe

The Planet Life team are in a small changing area putting on white lab coats, because that is what you do when you go in a laboratory, and they are about to go into a very special one. A sign on the door says, *Lab at the Beginning of the Universe.*

I do wish he would hurry up, thinks one of the Planet Life team looking at her watch. It is already 11.20 and I am meeting my friend for lunch at 12.00. At that moment a slightly flustered, out-of-breath Universal Builder arrives. 'I am very sorry that I am late, I was held up. Shall we go in?' he asks, grabbing a white coat, and with that they follow him through the door into the lab.

The first thing they notice is a huge, and rather battered, old wooden workbench upon which is a collection of laboratory equipment. Around the walls are shelves containing more equipment and in front of the workbench are wooden barstools, which they are ushered towards and asked to sit on. Once the team are perched on their stools, the presentation begins.

'The *Lab at the Beginning of the Universe* is where we do experiments on the Big Bang', announces their host looking for a reaction in his audience, who mostly look slightly perplexed, but he has their full attention. From the workbench he picks up a small container, and he brings it forward so that everyone can see.

The container is made of a mystery material, something like a type of rock. It hinges open to reveal two spherical chambers joined by a small channel to a third tiny chamber. They all peer inside. 'This chamber', he says pointing to the chamber on the left, 'collects the antimatter. And this one', pointing to the righthand chamber, 'collects the matter. The tiny one in the middle is for the nothing from which the matter and antimatter are made.'

'Is this how you created the matter and antimatter that your colleague showed us in the first presentation?' says one of the team.

'Yes, it is', comes the reply.

'Why is the *Nothing* chamber so much smaller than the other two chambers? Surely it should be twice as big,' asks another one.

Before the host can reply, her colleague interrupts. 'Didn't you learn anything from the first presentation? Matter and antimatter are made from nothing, and nothing takes up no space at all.'

Luckily, the questioner has become used to her colleague and does not take it personally.

'Now,' continues their host, closing the container so that they could read a label on its lid. 'This container is for making *up quarks*.'

'What are *up quarks*?' comes the inevitable reply.

'Well, everything in the universe is made from fundamental particles. Fundamental particles are the particles that can come out of the nothing as matter and antimatter. You see, not all types of particle can do this.'

He pauses, waiting for this to sink in, and then continues. 'When we build a universe, the first thing is to create all the fundamental particles in the right quantities.' He waves a hand in the general direction of the shelves where they see similar containers of different sizes. They presume that these were for making other types of fundamental particles.

He returns to the workbench and swaps the first container for a second one twice the size. They can see that this one has a label saying, *Down Quarks*.

'We use a mixture of up quarks and down quarks to make neutrons. A neutron is made from two down quarks and one up quark. That is why the *Down Quark* container is twice the size of the *Up Quark* container.'

Now, turning to address the smart one, he asks, 'Would you like to explain why we need neutrons?'

'Yes, I would!' she replies excitedly. 'The universe is built from atoms, and an atom is made from a neutron, a proton and an electron. And now, we have just learned that a neutron is made from two down quarks and one up quark,' she announced triumphantly in the way of one who loves understanding things and has now learned something new.

Her less quick colleague is thinking, I am glad they have put me on creating bees. This is all too much.

Their host bids them all to stand up and join him at the workbench. He takes the smaller *Up Quark* container and connects it to some piping. He then does the same for the larger *Down Quark* container. The piping goes in two directions, and they see that everything is colour coded. Containers on the left are yellow for antimatter and on the right blue from matter.

Looking to the left, they can see that tubes go from the *Up Quark* and *Down Quark* containers, so that they are mixed in another container.

'This yellow container is where antineutrons are created from antimatter up quarks and down quarks', explains their host pointing to the left side. Then, turning to face the right side, 'And this blue container is where neutrons are created from matter up quarks and down quarks.'

They can all see that clearly; it is quite simple.

The smart one asks, 'What about this extra blue stuff on the matter side? What is that for?'

'Well,' replies their host. 'We can actually make electrons and protons from neutrons.' He paused long enough for them to think about this before moving to his right and touching a glass vessel. 'This big glass sphere is for collecting the hydrogen gas that we have created.'

Now summarising, 'In the middle we have the nothing from which we make up quarks and down quarks, matter and antimatter versions of each. To the left we combine the antimatter up quarks and down quarks to make the antineutrons. To the right we combine the matter up quarks and down quarks to make neutrons. We then use these to make hydrogen.'

The audience are nodding in a way that indicates that they had understood.

'I am afraid,' continues their host. 'We cannot actually run any of the equipment at the moment. It can be a bit dangerous with extreme gravitational forces and all that. Anyway,' he says with a mischievous smile, 'How time flies when you are enjoying yourself!'

The team member who was worried about missing her lunch appointment glances at her watch. 'It is still 11.20! Or has my watch stopped?' she exclaims.

'This lab does funny things,' replies the host with a grin. 'There is no concept of time before the Big Bang.'

Atoms are made of neutrons, protons, and electrons. It has been discovered that a neutron is made from quarks, two of one type called 'down' quarks and one of another type called an 'up' quark.

You will remember that the only constraint on the universe design is that it must be made from nothing, and the mechanism for this is to have pairs of matter and antimatter that come out of the nothingness. Quarks are believed to be fundamental particles. This means that they come out of the nothingness with their antiquark mate.

There is no constraint on what types of fundamental particles come out of the nothingness – there does not have to be a single 'God Particle' – or on what quantities. If we were baking and needed twice as much flour as sugar, then we would buy twice as much flour as sugar. Logically, if twice as many *down quarks* are needed as *up quarks* then twice as many of them should come out of the nothingness with their antimatter friend.

So, where are we? We have the building materials for neutrons and antineutrons. Somehow these particles find each other in the right quantities – or they come out of the nothingness as a threesome, as that would be simplest – and combine to build the neutrons and antineutrons. This is done very early in the life of the universe.

At this point there should be all the neutrons and antineutrons needed for universe building.

Another, potentially significant thing, has been observed, a neutron can split into a proton and an electron. This is a bit strange, a neutron is built from three quarks, and then unbuilds itself into two different things, the electron and proton.

The early universe is believed to be 75% hydrogen and 25% helium. Hydrogen in its simplest form is a single proton at its centre with a single orbiting electron. Helium has two electrons and two protons plus two neutrons.

Electrons and protons are always in pairs for all the atoms, always in equal numbers. Would it not make sense to make the proton and electron pair from the neutron? Always in pairs and in the right place where they are needed. I say this because you can read how it is believed that the mechanism for building an atom is to capture free electrons that are whizzing around in space.

Here is a potentially simple design – and we like simple designs – in which neutrons are formed from quarks, and then some neutrons are converted into proton/electron pairs to build hydrogen atoms.

But there is something else interesting about this process, there is now a way for matter and antimatter to behave in different ways. Fundamental particles must have equal quantities of matter and antimatter. These potentially build equal quantities of neutrons and antineutrons. Now, however, the matter and antimatter can diverge. Some neutrons can be turned into protons and electrons, while antineutrons can stay as antineutrons. The universe does not want anti-gold, anti-iron, anti-oxygen, and so on, neither does it want these to build into anti-stars and anti-planets. If it does not want it, it will not have it, because everything is done for a reason.

We have reached the stage in our story where the initial universal building materials have been created. These are the smallest atoms. The bigger atoms will be built in a second stage that involves squashing these atoms together under great heat and force.

From our position 13.8 billion years after the Big Bang, we can see that the aim was to make separate galaxies containing separate stars, planet, and moons. How did the universe go from hydrogen and some helium into galaxies, stars and so on?

1.TEN

Flat Universe

One surprising fact is that the universe is believed to be flat. Here we are in our Milky Way galaxy and, it is perhaps surprising, that stars in other galaxies are all in the same plane as the Milky Way; there are no galaxies above, or below us.

It has also been observed that the other galaxies are all moving away from us. One description is, 'Imagine you draw dots on a balloon with a felt-tipped pen. Each dot represents a galaxy. As you blow up the balloon all the dots move away from each other equally.' This is what the expansion of the universe looks like.

Flying in the face of conventional science, I want to propose that the universe is not flat at all. The universe started as a sphere – this incredible pressure vessel that contained all the particles to stop them escaping – and the universe has remained a sphere that has expanded massively. The fact that the expansion looks like dots on the surface of a balloon is because it is literally dots on the surface of a balloon.

In the early life of the universe the aim of the sphere was to contain everything, otherwise particles would have zoomed off everywhere and would not have been able to combine into bigger particles and finally into atoms. At this point there was a cloud of gas because hydrogen and helium are gasses.

The next stage in the expansion of the universe was for the opposite reason; instead of keeping everything together, the aim was to pull everything apart to give space for the galaxies to exist on their own. Indeed, the continued expansion of the universe is to counteract the gravitational forces that exist between galaxies, even over huge distances, which would cause them to slide together. The current observed expansion is to ensure that everything stays apart.

Imagine the universe as a huge sphere with all the matter (and antimatter) contained within it. The point of expanding is to pull everything apart. However, if the universe expanded and all the stuff inside it just stayed in the middle, or just spread out a bit more, as would happen with gas inside a balloon, it would not achieve the desired effect.

What is needed is that the material clumps together into protogalaxies and these move outwards with the universe. We are all stuck on the inside of a giant balloon with nothing left towards the centre of the balloon – the point where the Big Bang started – and nothing above us because that is the nothingness of empty space.

Light from distant galaxies would appear to us to be travelling on a flat plane but would in fact follow the curvature of the inside of the balloon because it could not escape outside the universal sphere.

Clumping Hydrogen Clouds

Hydrogen is a gas so we can imagine quite easily how a cloud of hydrogen inside an expanding spherical universe would just expand to fill the additional space, it would not go on to form galaxies. Something is needed to cause everything to clump together into separate galaxy-sized balls.

Up until now we have concentrated on the matter and largely ignored the antimatter. One of the great mysteries is what happened to the antimatter, because there should be equal quantities of both normal matter and antimatter, it is just that we do not know where it is.

Some antimatter is in normal matter. This is how scientists at CERN have been able to do experiments with it, and, for example, a subatomic particle called a meson is built from a quark and an antiquark. But what happened to the rest of it? Everything is done for a reason, and in the simplest way, so antimatter is created for a reason and not just as a by-product of creating a universe from nothing.

What if it is the antimatter that makes the hydrogen clump together into the galaxies and stars?

As mentioned before, we could imagine that while neutrons are being built from quarks, in parallel antineutrons are being built from antiquarks. It would have a nice simple symmetry to it. It must be assumed that matter and antimatter do not annihilate each other in this early superheated environment; what would be the point of creating everything, only to have it disappear again?

Earlier we discussed the expansion of a proton and an electron into a hydrogen atom by using the analogy of basketballs on the beach. We noted that it would take more than two trillion basketballs to make a sphere with a two-mile radius. In other words, a hydrogen atom is expanded to two trillion times its original volume through this process.

Imagine that antineutrons do not expand into anti-hydrogen atoms. OK, some may do for whatever reason, but generally it is not desirable to build antimatter atoms that turn into antimatter stars and have the potential to cause havoc with their normal matter counterparts; and this has not happened as far as we know. Without the expansion into atoms, two trillion antimatter neutrons would take up the same space as one single hydrogen atom. Antimatter would be two trillion times as dense as hydrogen and have two trillion times the gravitational force.

Perhaps in this early 'soup' of particles, the neutrons are meeting up with their neutron friends and the antineutrons are meeting up with their antineutron friends. After all they have been locked down with their other halves for ever (or never).

Presumably, given long enough all the antineutrons would have found each other and clumped together into one single ball of incredible density – 2 trillion times the density of hydrogen – with the normal matter, a cloud of hydrogen, surrounding it.

However, each phase in universe creation is carefully timed. It is not given long enough to form a single ball, but rather trillions of smaller balls of varying sizes. As the universe expanded, these antineutron balls effectively stuck to the inside of the 'universal balloon' and, with their massive gravitational force, pulled with them a cloud of hydrogen, a bit like pulling off separate bits of candyfloss.

It is interesting to note that this process only works because matter has already done the first stage of creating hydrogen and helium, making it much less dense than the antimatter. Expand the matter first to make clumping work.

This could be the mechanism to create the desired clumping and it could explain the purpose of antimatter in all of this.

1.ELEVEN
Galaxies

We talked about how galaxies are observed to all be in one plane. It might also surprise you to know that galaxies are also essentially flat with everything orbiting in one plane. This could be a consequence of our protogalaxy expanding outwards stuck to the inside of the universal sphere.

Orbiting is what everything does. The moon orbits around the Earth, the planets orbit around the sun, and everything in the galaxy orbits around its centre.

A Galaxy (or Catherine Wheel!)

I would encourage you to watch a YouTube video of a Catherine Wheel firework. These are the fireworks that are attached to a frame of some sort via a central pin, and jets of firework shoot out causing the whole thing to spin. If you pause the video and see the pattern created by the jets, and then compare this with the pattern of stars in our type of galaxy, you will see a remarkable similarity.

In other words, it looks like our galaxy was formed by globules of material being fired out from a spinning centre. How could this happen?

If we go back to our antineutron ball surrounded by a cloud of hydrogen, our protogalaxy. It is reasonable to assume that the antineutron ball would be spinning because everything spins. The incredible gravitational force of the antimatter would pull in the hydrogen starting with what is closest to the centre. This would squash the hydrogen atoms together to form the bigger elements – lithium, beryllium, boron, carbon, and so on. Short range forces, perhaps because of the extreme temperatures, would keep the matter and the antimatter from combining. The forces would be so massive that the crust of normal matter would explode, and great globules of material would be fired out from the centre.

Something similar is observed in 'White Dwarfs'. These are stars that, when they cool, they can no longer support their own weight, so they essentially explode.

This would be a star manufacturing process. All the raw material in the form of the cloud of hydrogen (and some helium) that comes from the initial atom manufacturing process, is squashed to form the bigger elements, and then fired outwards to be delivered to their final destination. This goes on until all the material has been used up, and the galaxy has been formed.

1.TWELVE

Bigger Atom Manufacture

That is enough on galaxy creation. Let us now think about something a bit simpler: how are the bigger atoms manufactured from hydrogen and some helium?

In the beginning there was just hydrogen (and a bit of helium), so all the other elements were made from these two elements.

Helium

We need to remember our periodic table. Oxygen, for example, is number 8. Most oxygen is oxygen 16, meaning that it has 8 protons and 8 neutrons. Some oxygen is oxygen 18, meaning that it has two extra neutrons. It always has 8 protons otherwise it would not be oxygen. These are two isotopes of oxygen.

The starting point is hydrogen because that is all that existed, and the process of creating the bigger atoms is by squashing together the smaller ones.

How do we end up with oxygen 16? It could be 8 hydrogens [1], or 4 heliums [2], or 2 berylliums [4], or any other combination that adds up to 8.

ien	Oxygen 8	Fluorin 9
	O	F
I1	16.00	19.0
3.0	3.5	

There is one thing that does not stack up, however. Hydrogen is the only element without a neutron – it is just one proton and one electron – but oxygen has 8 neutrons, so if early hydrogen had no neutrons, how would we end up with 8 (or even 10) neutrons in oxygen?

Early hydrogen must have had at least one neutron to build bigger elements that have neutrons. Hydrogen with one neutron is an isotope called deuterium. Then two hydrogens would make one helium which typically has 2 neutrons, one hydrogen and one helium would make lithium with 3 neutrons, two heliums could join to make beryllium with 4 neutrons. It goes like this; the default is for each element to have the same number of neutrons as protons.

These elements can have isotopes with extra neutrons such as oxygen 18 as mentioned above, lithium 7, beryllium 9, boron 11, carbon 14, and so on. Where would the extra neutrons come from? Presumably from hydrogen as it all started there. Hydrogen also has an isotope called tritium that has 2 neutrons that could be the source of the extra ones needed for the isotopes.

Elements higher up the periodic table can have a greater number of extra neutrons in their isotopes than those near the beginning of the periodic table. Gold, number 79 in the periodic table, has an isotope gold 197, meaning it has 118 neutrons instead of 79. Presumably, this is the result of smaller elements with extra neutrons being squashed together, so the number of extra neutrons goes up.

Do you remember the alchemists who tried to make gold from lead? Smaller things get squashed together to make bigger things. Lead is number 82, and therefore bigger than gold, so how could gold be made from lead? Try making lead from gold instead by combining it with lithium [3].

Some isotopes are unstable, so we get radioactive decay. Carbon 14 for example, will eventually turn itself into carbon 12, meaning that it will lose its two extra neutrons. Did this decay happen to hydrogen too? Is this why most hydrogen is just a proton and an electron? Did hydrogen lose its early extra neutrons? Remember how a neutron can become a proton and electron pair, so maybe this process created more hydrogens.

Think also about the relative quantities. Gold [79] is scarce and iron [26] is abundant. Perhaps the different potential combinations of smaller elements to make a bigger element is important when creating more of one relative to another. Also, as the smaller ones are turned into the bigger ones, they are getting used up; it was all hydrogen to start with of course. It is worth noting that the overall relative quantities that are created are extremely important.

Gold
What if gold was in fact extremely common?

'What an awful bathroom!' whispered the wife to her husband in a conspiratorial way, after returning from upstairs, while their hosts were busy in the kitchen. 'Everything is 24 karat gold. Gold bath, gold sink, gold taps, everything! I mean how cheap!'

'Now, now dear,' replied the husband in hushed tones. 'Aren't you being just a little bit snobbish? Remember that Brian was made redundant when they were doing up their bathroom, so they had to go for the cheapest. I believe they went for B&Q Basics range, which obviously is made from 24 karat gold.'

'And,' continued the wife unperturbed. 'Did you see Brian's gold bike lying on the grass in the front garden when we came in?'

'Well, a gold bike is a good idea dear,' replied the husband. 'You can leave it anywhere without being worried that it will be stolen.'

We used the word 'squashing' for combining smaller atoms into bigger ones. This really does not do justice to how much force is required. In the Tesla car factory, sheet metal stamping presses squash the metal under a staggering 4,500 tons of force. Two pieces of aluminium [13] in a press will not squash into a piece of iron [26]. The force and heat required for bigger atom manufacturing must be truly incredible.

1.THIRTEEN

Stars

We previously discussed how galaxies formed. Now let us consider how stars formed within galaxies.

As the universe expanded, everything moved outwards squashed up against the inside of the universe, and this could account for why everything in a galaxy is in one plane, and why all the galaxies are also in one apparent plane.

We suggested that balls of antineutrons, with their extreme gravitational force, pulled a cloud of hydrogen gas around them to cause the gas to be pulled apart into separate galaxies. The extreme gravitational force of the biggest ball of antineutrons spun around, squashing the hydrogen, and firing balls of material away from the centre.

Now, if we think about it, why would there only be one ball of antineutrons? Surely there would have been many balls of different sizes, which all attracted clouds of hydrogen. These ones could have formed stars.

It has been observed how giant stars will eventually cool and collapse as they are no longer able to support themselves. Some of these collapse into Black Holes.

If we think of a volcano exploding, the molten lava becomes solid when it cools. Solid stuff is stronger than liquid stuff, so why would a star that became cooler and, therefore more solid, collapse?

A possible explanation is that stars formed around a ball of antineutrons. The gravitational forces exerted by this ball were big enough to squash hydrogen into bigger elements, but not so big that the star collapsed. The early universe was incredibly hot and under huge pressure, and in this environment matter and antimatter did not recombine. The centre of a star is also incredibly hot and under huge pressure, so it is possible that antimatter in the centre of the star is stable.

However, when the star has cooled through doing what a star does, which it is to radiate heat, the antimatter could become dangerous. Antimatter and matter at the core of the star start to annihilate each other. The star would be eaten away from inside and this would cause the star to collapse.

The Sun

Our star, the Sun, is another interesting case, because it is a huge ball of gas which is wider at its equator than at its poles as it spins. The Sun is one million times the volume of the Earth. It is so huge, but it is still gas. Why is it still gas? It appears that it does not have the gravitational force needed to crush hydrogen and form the bigger atoms. What is going on?

If we think of raindrops forming in a cloud, they only form when there is something for them to form around, indeed artificial rain can be made by flying over a cloud and dropping silver iodide crystals to seed the rain drops.

It would seem likely that seeding would be required to cause a ball of gas like our Sun to form. Could it be that a star like our Sun has a smaller ball of antineutrons at its core? This caused the Sun to form but did not have the gravitational force to squash hydrogen into bigger atoms.

A Black Hole

The centre of a galaxy is observed to be a giant Black Hole. This could be because the huge antineutron ball has done its job of manufacturing the bigger elements and firing out great globules that formed planets and moons, and maybe some stars. It now has the job of keeping the whole galaxy together through its extreme gravitational force as a Black Hole. A Black Hole in the centre of galaxies and inside some stars. ·

It is worth noting that observations have shown that there is a mismatch between the observed amount of matter and the gravitational forces. There is too much gravity for the matter. Because of this, 'Dark Matter' has been proposed. This is additional matter that cannot be observed but must be there for the gravity calculations to be correct.

Remember that there is also the mystery of what happened to the antimatter. Could it be that the missing gravity and the missing antimatter are down to antimatter lurking in Black Holes?

1.FOURTEEN
Our Planet

Our planet comprising solid rocks and all the nice atoms that we need, now combined into molecules – calcium carbonate, iron oxide, and so on – could have started life as a globule fired out from the centre of the galaxy.

The early galaxy was like a giant pinball machine. Stars like our sun were already out there; the Sun is 10 million years older than the Earth. A globule like the Earth was on a perilous journey. It could have whacked into something just like meteorites that crash into Earth, but it was on the perfect trajectory that took it past the Sun at exactly the right distance.

It became caught up in the Sun's gravitational field, causing it to divert into an elliptical orbit that eventually became circular.

To be a planet capable of supporting life, it must have passed by the Sun at exactly the right distance to be in the 'Goldilocks Zone' where water is liquid. Then there are all its other characteristics such as its size that provided the gravity we experience, rotating once every 24 hours, and orbiting the Sun once every 365 days. It all adds up to a rare combination.

What about our atmosphere? It is believed that the Earth formed its own atmosphere. There were many volcanoes in the early Earth as it cooled, and the Earth's crust formed. These would have released steam (H_2O), carbon dioxide (CO_2), and ammonia (NH_3). Ammonia would have been broken down by the Sun's radiation to go back to nitrogen and hydrogen.

1.FIFTEEN
Recycling and Disposal

Let us turn our attention to what happens next.

The first thing is recycling. You can read how stars are formed from clouds of dust and debris being squashed together. This would be a secondary process because dust and debris did not exist in the early universe, just clouds of hydrogen. It would imply that there is recycling going on. A star explodes – as has been observed in White Dwarfs – and this material recombines into new stars.

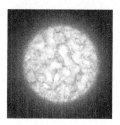

White Dwarf

The final act will be disposal. The universe started as nothing, and we can be sure that it will eventually return to nothing. How would a universe go back to nothing? Surely the only way is the reverse of the way that it was created. Matter and antimatter must eventually recombine to annihilate each other.

The galaxy design that we have discussed could also provide a means of its disposal. If antimatter is lurking in Black Holes, eventually the matter can be sucked into those Black Holes and annihilated. Big stars have their own built-in Black Holes, while others will be sucked towards the Black Hole at the centre of the galaxy.

Interestingly, experiments show gamma radiation given off when matter and antimatter annihilate each other, and Black Holes are also observed to give off gamma radiation.

We noted earlier how the universe cannot be traced back to exactly the time of the Big Bang because time did not exist at that point. Similarly, it has been observed that the laws of physics break down once you go over the event horizon of a Black Hole. Could this be an indication of stars going back to nothing?

We have also discussed how the universe itself expands in the way that it needs to at each epoch of its lifecycle. It can be imagined that the universe can also shrink when the time comes for it to die. Perhaps the large chunks of matter are disposed of as described above, and then the universe will shrink, and, like a giant fishing net, other matter and antimatter will be brought together. This will continue until the universe itself has shrunk back to a singularity.

Little Bang, and the universe is gone. There is no sign that it ever existed.

1.SIXTEEN

Infinity

Finally, let us think about infinity, something that we cannot truly understand.

At the beginning of this section, we talked about how the space in which the universe is created is infinite. Mathematicians can get excited about what is meant by infinity. If you came up with the largest number you could think of in the trillions and quintillions, someone could always add another number on the end of your number. There is always a bigger number, an infinite number of numbers.

What about small numbers? You could think up an extremely small number, zero-point-zero-zero-zero-zero and so on, but someone could always make a smaller number than your one by adding another zero after the decimal point. Infinity can be infinitely small as well as infinitely big.

The nothingness in which, and from which, the universe is created has no size, it is both infinitely big and infinitely small. Infinity and nothingness are two ways of describing the same thing. The universe is not infinite – that is the point – but it started out at the infinitely small end of the spectrum and has grown towards the infinitely big end.

When we say that matter and antimatter annihilate each other and disappear, perhaps they are not just annihilating in a zapping sort of way. Is it not a process in which they are losing their size and becoming infinitely small?

Somehow structure has been created where there was none. This is the fundamental mystery in universe creation.

We take everything for granted, the sun and moon, the days and nights, the seasons, the years. It is our reality. We don't need to question it, but, if we do, we find a unified design that we can understand at some level.

Reality is a fragile thing. We know that when we lose someone, or something, we took for granted. When people come up with fake news, deep fake images, virtual reality computer worlds, and so on, it can be deeply disturbing. Reality is really important.

1.SEVENTEEN

The Summary

The Final Talk

The Planet Life team are gathered for one final talk from Universal Building to summarise what they have learned about making a universe.

The presenter is saying, 'Let's recap all the stages in making a universe. It will be useful in case any of you are transferred to our department.'

They are all thinking, no thank you, I am quite happy where I am doing planet life, all except the smart one who had found the presentations really interesting. It has to be said that her colleagues had noticed a change in her from being quiet and shy at the beginning of the sequence of talks, through to increasing levels of confidence, some would say over-confidence.

The presenter starts off by asking, 'What is the starting point for a universe?'

The smart one puts up her hand and, without first being asked, says, 'Nothing! Endless supplies of nothing.' Then she continues knowledgably, 'Endless nothing is good for when the universe expands and gets really big but is not necessary for making the universe because nothing takes up no space at all.'

'Yes,' replies the presenter. 'Well put. What next?'

'The Big Bang!' replies the whole group trying not to laugh. They are in that sort of end-of-term mood.

When the group has settled down, the presenter enquires, 'Is the Big Bang done directly in the nothing of empty space?'

'No!' comes the confident reply. 'If that happened, all the particles would zoom off in different directions. They need to be contained in a tiny universe.'

'That is right. The Big Bang releases fundamental particles. Equal quantities of matter and antimatter ones. We must combine them to make other things, so if they all zoomed off everywhere this would not be possible.'

Then he asks, scanning the different faces, 'Can anyone remember, where does energy come from and where does matter come from?'

Most of the group could not remember that one, but the smart one could. 'They both come from the fundamental particles. They are the same thing. Energy is matter and matter is energy.'

The rest of the group were looking impressed. 'Yes, well done,' says the presenter.

Next, he continues, 'You will remember from when we were in the *Lab at the Beginning of the Universe*, how we make neutrons and antineutrons from quarks and antiquarks, and then we make hydrogen from the neutrons?' They were all nodding as that had been a memorable visit to the lab. 'When a neutron splits into a proton and electron to make hydrogen, how much does that pump up the volume?'

After a pause, 'Was it 2 trillion times?' comes a hesitant voice.

'Yes, it was. That is a lot isn't it? Can you remember how we do that?'

Even the smart one wasn't sure, so he continues, 'The electron orbits the proton, but it is not like the Moon orbiting your Earth. That doesn't make the Earth bigger does it? It is this clever idea that the electron is everywhere at the same time, so it makes the surface of a bigger thing from an extremely tiny thing.'

Most of them are turning to a neighbour shrugging and putting on puzzled expressions, as they don't remember that important bit of the process.

'OK,' he continues. 'You will remember that we get to the stage of having a mixture of antineutrons that contain the antimatter fundamental particles, and hydrogen that is built from the matter fundamental particles. We have expanded the universe massively because we have made enough stuff to build everything in the whole universe, and the matter has pumped up 2 trillion times the original volume.

'Can anyone remember what we do next?' he asks.

At this point, the smart one starts waving her hand in the air, and jumps in before anyone else has a chance, somehow imbued with an extra amount of keenness, 'You keep expanding the universe,' and, before what would undoubtably be the next question, she continues without pausing for breath, 'Balls of antineutrons with their huge gravitational force which is 2 trillion times that of hydrogen attract clouds of hydrogen around them and everything travels outwards with the expanding universe until there are the beginnings of separate galaxies.

Her colleagues are making, 'What's up with her?' type of faces to each other and trying not to get hysterics. That would be inappropriate in a meeting, making it all the more likely to happen.

'Precisely,' replies the presenter with a hint of a smile as he was having the same problem, and desperately trying to keep a straight face, he asks, 'What do we do next?'

Although this was not funny at all, everybody falls about laughing, leaving the smart one wondering what is going on with everyone else.

The moment passes, and the group all noticeably lean forward now in serious, concentrated mode to mitigate against any repeat of the hysterics.

'Would you like to continue?' asks the presenter raising an enquiring eyebrow to the smart one, hoping that this will not cause a repeat of what has just happened.

'Yes!' she replies unnecessarily keenly. 'The biggest spinning ball of antineutrons squashes hydrogen to make the other elements.'

Then, she stands up so that she can do the actions more easily. 'It spins around like this and flings out great globules of stuff in all directions,' she says, combining flinging actions with the spinning.

Did she do that deliberately? wonders the presenter wiping a tear from his eye, with his audience creased up in fits of laughter.

Now with no chance of continuing with any serious discussion, he says to anyone who is still listening, 'That was a wonderful end to our inter-departmental talks. I have thoroughly enjoyed explaining what we do in Universal Building and look forward to coming to learn about what you do in Planet Life.'

There is a spontaneous round of applause from a happy group. The smart one is beaming after the unexpected triumph of her demonstration.

SECTION 2 – PLANET LIFE

How to Make Life of Earth

In the previous section we discussed how the universe was built from nothing. Our tiny planet – the third rock from our sun – was created through this process: a small, lifeless ball of rock.

Now let us look at Life. This is a bit easier as we are on more familiar territory. Let us consider how Life went from nothing to all the incredible, varied, and beautiful lifeforms that make up our ecosystem.

2.ONE

The Beginning

There they were, Adam and Eve in their Eden garden. We know there was an apple tree, a fig tree, and a snake. There must have been other things too, bearing in mind that apples – and snake - were off the menu, a diet of figs every day would not have been the dream life that we were told it was.

Presumably, a garden would have had flowers, and then bees or other insects would pollinate the fruit trees, and there would need to be soil, and worms aerating the soil, and oxygen to breathe, and carbon dioxide for the plants, and an atmosphere with a good ozone layer to stop the naked people from being burned. The list goes on because there needs to be a complete ecosystem.

We have only survived outside the Earth's ecosystem in spacecraft, and these are always re-supplied from Mother Earth. Say we tried to colonise Mars, what would we do? Presumably build some sort of bio-dome and then try and create a self-sustaining ecosystem that could support life. How would we do it? By copying how it is done on Earth, only our version would be totally rubbish in comparison. There would be the rocky inhospitable surface of Mars and its noxious atmosphere, and somewhere there would be a dome like the Eden Project containing plants and everything needed that had been brought from Earth.

Our creator turned the whole planet into a self-contained ecosystem. We can look at our Moon, or Mars or other planets and get a good idea of what a planet looks like when there is no life. And just for the record, without being Marsist, Mars has had just as long as Earth to get life going, so should we find a tiny bit of bacteria lurking under some Martian rock, that does not count in my book.

How do you turn an inhospitable planet into one teaming with life? The key thing is location, location, location, something that Mars and the other planets do not have. Then it is Life itself that turns that inhospitable environment into a place where it can thrive.

Life always had the aim to create ever more complex versions of itself, and these had no chance of surviving in that early environment. So, Life had a long-term plan to get to where it wanted through a sequence of steps.

The Start of Life
How did life on Earth get started?

One theory about how life started on Earth was that it came on a comet from outer space. Where did this comet land? It could have landed anywhere so let us pick a location at random.

UK readers will know that when you travel west from London to Wales, you take the M4 motorway, and about halfway along you get to the Membury Services. Let us say that our comet landed billions of years ago at what is now the site of Membury Services.

A Comet

This means that modern day visitors to the Services, would note a plaque on the wall that says, 'This is believed to the be the site where a comet landed bringing with it the first life on Earth.'

Intrigued by this, the curious visitor can read on, 'After 100 years Life had reached Bath to the west, Winchester to the south, Oxford to the north, and Slough to the east. Unfortunately, at this time a huge volcano erupted somewhere up the A34 near Oxford. The ensuing lava and volcanic ash wiped out all of Life except a small pocket outside Slough.'

Life thought, Uh oh! I didn't even make it as far as Wales, let alone to any of those other countries that I saw as I orbited the Earth a few times before making a perfect landing at Membury. I am going to have to start all over again.'

Another theory is that it could have started around super-heated water vents deep in the ocean. Alright, the conditions might be right for life to start here, but it seems a bit random.

Life on Earth is not a random accident, it is a major project, and like all good projects it will have been meticulously planned.

The first step in creating life on Earth is perhaps the most critical. It is like the 'D' Day landings; the aim is to establish a Beach Head. There will be a lot of casualties, but if the first step fails there cannot be any further steps.

What would we do? How would we give ourselves the best chance of success?

I suggest that Life should start in the atmosphere. Imagine trillions of miniscule spores raining down from the sky. Some would fall on rocky ground, which is fine because the only ground is rocky; earthy ground is created by years of rotted vegetation. Some would fall in the sea, which is good too, in some ways it is better because the sea protects this early life from the radiation, meaning that life in the sea can develop faster than life on land.

Rather than a single comet carrying life from a mystery life source that lands in one place at just one time, this first life could go on raining down all over the Earth for years if necessary, go on until it is sure that it has established itself. That is a much more certain way of doing it.

2.TWO

The Simplest Life

The birds and the bees, and the flowers and the trees. These are examples of life as we know it. Clearly, the first life would not have been like this, it would have been the simplest organisms that could survive and reproduce.

Organic structures of any complexity are produced by organisms; organisms are factories that can grow and produce other versions of themselves. As they become more sophisticated, they can build more sophisticated organisms.

This gives scientists a puzzle because there were no organisms to create the first organisms. The first life had to be created out in the open. This life would have been simple by organic standards, but massively complex in non-organic terms.

Would it have been possible to create organic structures out in the open, up in the atmosphere as we have suggested? The simple answer is yes.

We must remember that the early atmosphere of the planet would have been horrible. Volcanoes erupting, potentially much hotter, no ozone to protect from the sun's rays. In fact, it could have been like a laboratory, perfect for creating complex 'organic' molecules, something that would be impossible now.

Recently phosphine gas was observed in the atmosphere of Venus. Scientists were excited because, phosphine is known to be created by organisms, not out in the open in a planet atmosphere. They thought, that should not be there, that is not made using normal chemistry. It must have been made by an organism. Maybe there is life on Venus?

Alternatively, it could be an example of a compound that we consider to be organic, existing in the atmosphere outside of any organism.

Early Lifeform
What might the earliest simple life have been like? We do not actually know because this was over 3.5 billion years ago, but perhaps bacteria spores and seeds can provide an insight.

Some bacteria, sensing impending doom, go into lockdown mode. An emergency sequence of genetic instructions turns them into a spore that preserves their DNA within an indestructible shell. This comes in the 0.1% of bacteria that well-known disinfectants cannot kill, along with being immune from the effects of extreme heat and radiation.

Seeds are similar in that they have their DNA inside a shell that also contains just enough nutrients to get life started.

These are simple organic structures that are not alive, but they are not dead either. They can become alive when they sense that the conditions are right. This would be the simplest type of structure that could start life on Earth.

Scientists from California have revived ancient spores and got them going again.

A Bee in Amber

Bacteria Spores

'Isn't that one of those 25-million-year-old bees trapped in amber?'

'Oh yes, so it is, let's take it to the lab and see if there are any 25-million-year-old bacteria spores in its stomach.'

'Yes, there are some. Shall we see if we can get them to them to come alive?'

'What could possibly go wrong?'

'No. Cannot think of anything. By the way did you see Jurassic Park the other night?'

So, they did, and microbes multiplied in their culture dishes.

Culture Dish

Isn't this strange? Something can be 'not alive' for 25 million years or longer, but not dead either. When it senses the right conditions, it starts operating. This is down to the secret magic that makes life happen. Let us hazard a guess that we will never know what this is and let us hope that we never know, or we will 'play God' even more than we do already.

We can imagine that the creation of this early life would be triggered at a specific time when Life sensed that the atmosphere had calmed enough that it could survive. We see triggers all over the place such as a solitary locust deciding to join a swarm, or Mayflies all hatching and mating in a single day.

Spontaneously trillions of Life molecules would have formed in the atmosphere and rained down on the land and the seas, becoming alive when they sensed that the conditions were right.

Genetic Code Molecules
We talk about DNA meaning the genetic code. Simpler types of molecules would have been used instead of DNA for early life. RNA has been suggested. Our bodies have DNA to store the code and RNA to read the code, but RNA can be used on its own. Other types of molecule have also been proposed.

These genetic code molecules would build themselves, then more molecules would have been created around them for the nutrients and the shell.

The shells of bacteria spores are made of a type of protein. All these fancy names for organic structures, but underneath they are mainly hydrogen, nitrogen, carbon, and oxygen. RNA and DNA also use phosphorus for their backbone – phosphorus and phosphine? Some proteins contain sulphur. These are all available in the early atmosphere.

Think how small they would have been, much smaller than a bacterium, which is only 5 micrometres across, or a coronavirus, which is even smaller at 0.12 micrometres.

Activating Life

Once on the ground, or in the sea, they would need to start operating, just as seeds or spores start germinating when they sense that conditions are right to do so. What would operating mean?

The first aim was to become alive and stay alive, to 'metabolise'; doing what it was programmed to do. The most basic thing it must do is grow and then make a copy of itself to reproduce. How would this happen? We know from our own machines that they need power. The earliest organisms would get their power from the sun through photosynthesis.

Photosynthesis takes in water and carbon dioxide and excretes oxygen. It does this primarily to build cellulose from which all plants are made.

Would this work in the sea as well as on land? Yes, because there is plenty of water in the sea, as well as dissolved carbon dioxide.

However, it is not enough for our creator to go, 'Great look it is working!' This earliest life is there for a purpose, firstly to survive, but then to provide the little factories that will allow more complex organisms to be created, because these could not be created out in the open.

There was a second mission too, to start the process of creating an environment in which more complex life could survive. It would do this step by step over a long period.

It is fascinating to think that the genetic code for this earliest life – the pattern of atoms in the code molecule – must have been designed before any of this started, and the plan of how to go from this first step onto the next steps was already in place.

We can also say that the event that triggered life on our planet would happen on every planet capable of supporting life. There would be little point in creating a life-sustaining planet and then not starting life.

It also makes sense that life must start as early as possible so that it has the longest time in which to evolve and survive. It would be pointless to get only as far as dinosaurs when the Sun has burned itself out, which could happen if life had been brought in on a random comet.

Life is part of the lifecycle of the planet, and living organisms change its environment. It is not Life *on* Earth, rather it is Life *of* Earth.

2.THREE
Organic Chemistry

Let us think about this organic chemistry some more.

Do you remember your chemistry? Hydrogen – poor hydrogen – being electronly-challenged with its one electron, which obviously is not enough, wants to bond with another element. It could find another one of itself to make H_2, alternatively two of it could bond with oxygen because oxygen – number 8 in the periodic table – has two free electrons. Whereas helium, at number 2 in the periodic table has, in a self-satisfied way, got exactly the right number of electrons, so it does not need to bond with anything else.

It goes along like this in a sensible, mechanical way. Na and Cl will bond together, iron oxide, aka rust, calcium carbonate, carbon dioxide, and so on.

Nobody taught me about the chemistry going on in living things, organic chemistry. It is like, forget all these boring rules, we will tear up the rule book. If you want to make something with 54 carbon atoms, 108 hydrogen atoms, and 6 oxygen atoms, knock yourself out, go for it. That is fat by the way. What about having 38 carbons, 64 hydrogens and 4 oxygens to give us omega 6 fatty acid? Or 60 carbons, 92 hydrogens and 4 oxygens for omega three? Or 38 carbons, 64 hydrogens and 4 oxygens for starch?

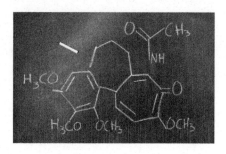

I have a vision of the Planet Life* team sitting in a big room with blackboards, drawing out triangles and squares and hexagons with C and H and O written in the corners to create different structures. One of them calls over a colleague, 'Look I have made $C_{54}H_{108}O_6$ – it is this wobbly, blobby stuff, but I cannot see any use for it.' But then the colleague goes, 'I could use that for storage around the stomachs of my creatures. I was going to use $C_6H_{10}O_5$ [cellulose] which we use for making tree trunks, but that might have restricted their movements somewhat.'

I am sure that someone will tell me that underneath it is the same sensible, mechanical bonding going on as in normal inorganic chemistry, but it is a nice picture.

*For those who have not yet read the section on the Universe, we are imagining – not seriously – that the Creator has teams working on the creation tasks. There is the Universal Building department who build universes, and the Planet Life department who do life on planets.

Complex compounds are called organic compounds and are made inside organisms. Our chemistry is inorganic, we find it hard to do the complex stuff. If we want to build something with wood, we use real wood made by nature. If we make wood, it is plastic wood not real wood; or fertiliser, we make inorganic nitrate fertilisers, if you want organic go straight to the horse or cow.

All living things, including us, are factories, or chemical laboratories, that build molecules from other molecules, and are constantly working at atomic level.

In a Pilates class the instructor might say, 'Breath in…. breath out' and you think, I was planning on doing that anyway. I will breathe in oxygen, do some processing of the molecules, and breathe out carbon dioxide. It is the same for 'Eat in… poo out' and 'Drink in… pee out', only not normally during a Pilates class.

Fat Belly

'Excuse me,' a very thin woman says approaching a man in the park. 'I was rather admiring your fat belly. From the side you can see that it hangs down over the top of your trousers and you have wrapped your t-shirt underneath and stuffed it in.'

'Thank you very much for the compliment,' beams the man back and continues. 'Yes, I am very proud of that, I have created it with atoms.'

And noting her enquiring expression, he continues, 'Yes, I consume C_2H_5OH [alcohol] in the form of large quantities of beer, and then I move the atoms around to create $C_{54}H_{108}O_6$ [fat] which I store in my stomach.'

A large woman, who she assumes is his wife, turns to the thin woman and says, 'I know it is not PC, so I hope you won't find me 'thinnest,' but you are really skinny. There is hardly any fat on you at all!'

'Well,' says the thin woman. 'I hardly ever drink C_2H_5OH [alcohol] and I try to cut down on things with $C_{12}H_{22}O_{11}$ [sugar], and I am mostly vegetarian.'

She observes their open mouths, slightly turned down like Wallace from Wallace and Gromit. They appear to be becoming unsettled.

'Yes,' she continues. 'I eat the leaves of vegetables like cabbage and lettuce, and the roots of vegetables like carrots.'

She becomes aware of a secret game of statues as the couple creep ever closer to each other until they are squashed up against each other holding hands for moral support.

Undeterred, she continues, 'Sometimes I have pulses – beans, you know – and I eat fruit such as apples and lemons.'

The couple's expressions have now gone to the 'This is weird, we've gotta get out of here' stage.

But she continues, 'Cabbage for example is a good source of $C_{31}H_{46}O_2$ [vitamin k], lemons have $C_6H_8O_6$ [vitamin C], beans contain

$C_m(H_2O)_n$ [carbohydrates] and...' At this point she senses she has said enough.

There is a marked sense of relief in the couple's eyes, and the woman starts speaking with an air of concern and pity. 'So, you are getting all the atoms that you need for a balanced diet, but you just don't know how to convert them into $C_{54}H_{108}O_6$ [fat]. So sad.'

'Anyway, nice to talk to you, we must be off now,' says the man. 'Goodbye.'

Vitamin C

Living organisms are mostly made of the same ingredients, carbon, hydrogen, oxygen, and nitrogen. These are available in all good planet atmospheres. There is no need to keep bringing in life on a comet from outer space when everything you need is right here.

It makes sense that the simplest forms of life would be made from the most easily available materials. We see that in things we make ourselves. The Flintstones made everything from stone, indigenous people will make their houses from trees and branches. Then, as we make more sophisticated things, we need more exotic materials. It is the same for our bodies, 'Have you had your magnesium? You

106

are looking a bit anaemic, you need iron. Your legs are too bendy due to calcium deficiency.'

Pigeons are very keen on eating cabbages that people grow in their allotments. We consider cabbage to be rather boring, but it is full of nutrients including vitamin C, potassium, magnesium, and vitamins A and K. Clever pigeons. Well not really, we can assume that, like us, they have no clue about what is going on. It is Life itself that knows the materials it needs to build and repair itself.

Corn Starch and Syrup

A Sidenote

When we think about how our bodies turn molecules into other molecules, it brings into focus the potential dangers of messing around with this process in our food – as well as in animal and plant feed – by changing molecular structures in our industrial processes, corn starch $(C_6H_{10}O_5)_n$ converted into corn syrup $C_6H_{14}O_7$ is an example. It might be alright, but it might not. Machines run better on the fuels they were designed to use, best to have food as nature intended and everything in moderation.

2.FOUR

Single Cell to Multi-Cell

Life had a plan from the start to become more complex and more diverse. There are an estimated 8.7 million species on our planet: us, dogs, fluffy white rabbits, blotched blue tongued skinks, spiny lumpsuckers, pink fairy armadillos, raspberry crazy ants, tasselled wobbegongs, hellbenders, sneezewort, hooded skullcaps, yellow toadflax, and skunk cabbage, to name just a few (well, sort of).

It is important to understand that these are not separate things, rather we are all part of one thing called Life, the Life of Earth. Darwin was ridiculed for saying that we were related to apes. That was before it was known that we share 99% of our DNA with chimpanzees. What fun the cartoonists would have had if Darwin had known we share 60% of our DNA with a banana, though at 80% a potato is a closer relative.

Of the 8.7 million species, our version of Life is the best (according to us at least), oh yes, we are even made in God's image, so you cannot get any better than that*.

*For those not in tune with the British sense of irony, this is a view that may not be shared by, well, most of the other lifeforms. The orphaned orangutan whose rainforest has been cut down, the last remaining rhinos dying so that their horn can improve someone's sexual performance, and so on.

When we think about the earliest life – the single-celled organisms – it is more than a passing interest. These are our ancestors; without them we would not even exist. Let us return to the earliest life on our planet which was believed to have started 3.5 billion years ago. The universe is 13.8 billion years old, the Sun 4.6 billion, and Earth 4.5 billion, so the Creator was not in any great hurry.

A single-celled organism takes in nutrients and ejects waste through its cell wall. We eat, drink, and breathe and do those other things on our big scale, and cells do the same thing on their tiny scale. Tiny, as we have discussed, means working with atoms. It is quite remarkable when you think about it, something so small and simple stripping off atoms from molecules, using them to build other molecules, and ejecting waste atoms and molecules. It is a nice simple design; chemists can work out quite easily what inputs are needed to create each bit of the cell.

A photosynthesising single-celled organism takes in its H_2O and CO_2 and phosphorus (probably) for its 'DNA' and ejects oxygen. It does this because it is following the instructions in its DNA. It keeps on expanding until it is double the size.

The DNA also instructs it to build an exact copy of its DNA. The two DNAs then move left and right within the cell and the cell splits down the middle to become two organisms with one copy of DNA each. The two single-celled organisms now repeat the same set of instructions dictated by their identical DNAs.

This is a simple lifecycle that involves just building and replicating and then going again.

Multicellular organisms use the same process, except that their cells remain joined. We have seen those slowed down films of cells dividing, looking very much like frogspawn in a pond. Because it is the same process as for single-celled organisms, the DNA is copied to each cell even though only a part of the code is needed for building that part of the organism.

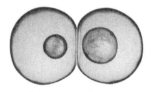

Cells Dividing

The earliest multicellular organisms are believed to be cyanobacteria that date to around 2.3 billion years ago. How did life go from its initial single-celled versions to multicellular?

We know that organic things of any complexity can only be created inside organisms. We also know the mechanism through which the single-celled organisms replicate is by making an exact copy of their DNA. Surely, that means that the only way to go from single-celled to multicellular is for the single-celled organism to create a different version of the DNA.

Instead of creating an exact copy of its DNA, and replicating to create a clone of itself, the single-celled organism would have created a new DNA with a different set of instructions.

When the cells divided, the new organism, with its new DNA, would have followed the new instructions, which this time would tell it to make a simple multicellular cyanobacterium.

When this one reproduced, it would have reverted to normal mode making an exact copy of its DNA with instructions to make another multicellular cyanobacterium.

This is an example of a stage in the lifecycle of Life. Life decided it that it was ready to go to the next level of complexity and it triggered a code change. Life has its lifecycle and each organism created by Life has its own lifecycle. There is no way that a single-celled organism, or indeed a multicellular one like us, is clever enough to be able to make this change themselves.

Just as we commented earlier that the code for the first life was already designed, the code to go from single-celled to multicellular must have already existed, and the lifecycle of Life must already have been planned.

2.FIVE

The Chronology of Life

A summary of the chronology of life (mostly taken from Wikipedia) is as follows:

- The age of the Earth is about **4.54 billion years**. **Carbon dioxide** (CO_2) has been present in the atmosphere since the Earth condensed from a ball of hot gases that exploded out of a star. The early atmosphere was mostly carbon dioxide, with little or no oxygen. There were smaller proportions of water vapour, ammonia, and methane. As the Earth cooled down, most of the water vapour condensed and formed the oceans.

- Early in the Earth's history—about 3.5 billion years ago—there was **1,000 times as much methane** in the atmosphere as there is now. The earliest methane was released into the atmosphere by volcanic activity. During this time, Earth's earliest life appeared.

- As early as **3.5 billion years ago**, **bacteria** began to produce **oxygen** as a waste product of their activity.

- The **Great Oxidation Event** was triggered by cyanobacteria producing the oxygen which developed into **multicellular forms as early as 2.3 billion years ago**.

- The first **fish** appeared around **530 million years ago** and then underwent a long period of evolution so that, today, they are by far the most diverse group of vertebrates.

- **Insects** likely originated about **479 million years ago**. Insect flight emerged around **406 million years ago**, around the same time plants began to really diversify on land and grow upward into forests.

- The first **large trees** began to appear on Earth almost **400 million years ago.**

- The earliest known **reptile** is Hylonomus lyelli. It is also the first animal known to have fully adapted to life on land. Hylonomus lived about **315 million years ago.**

- First **Dinosaurs** appeared approximately **230 million years ago**, during the Triassic Period, the dinosaurs appeared, evolved from the reptiles.

- The earliest known **mammals** were the morganucodontids, tiny shrew-size creatures that lived in the shadows of the dinosaurs **210 million years ago.** They were one of several different mammal lineages that emerged around that time. All living mammals today, including us, descend from the one line that survived.

- **Flowers** and flowering trees began changing the way the world looked almost as soon as they appeared on Earth about **130 million years ago.**

- Fossil records suggest that modern birds originated **60 million years ago** when dinosaurs died off.

- The earliest firm records of **grasses** are from the Palaeocene of South America and Africa, between **60 and 55 million years ago**.

- On the biggest steps in early human evolution scientists agree. The first **human** ancestors appeared between five million and **seven million years ago**, probably when some apelike creatures in Africa began to walk habitually on two legs. They were flaking **crude stone tools** by **2.5 million years ago.**

The list shows how Life had a plan to introduce ever more complex and varied versions of itself. Single-celled creatures became multicellular. Plants developed into big trees and different species. Animals came along, starting with insects, then reptiles and finally mammals. With mammal babies supplied directly by nutrients from the mother, the most complex creatures could be created. It seems that mammals were the final level of complexity that Life wanted to reach.

Clearly there is no way that we could have survived in the early Earth. We breathe oxygen for a start and there was hardly any of that, and we eat plants, or we eat creatures that have eaten plants, so that would not have worked either. We are made from many complex materials, so that was a non-starter too.

Life had to take things in stages and could only do limited jumps in complexity from one lifeform to another. Earliest life had the role of getting established, but also of modifying the atmosphere and the environment.

Oxygen breathing creatures could not be introduced until there was sufficient oxygen in the atmosphere, after the 'Great Oxygenating Event'. Although, we think of plants creating the oxygen, this was largely down to the photosynthesising cyanobacteria in the sea. Volcanoes put phosphorous into the water and the bacteria thought, great I need that for building my DNA.

Oxygen had the effect of reducing the methane in the atmosphere, turning it into carbon dioxide. This in turn cooled the planet. Having oxygen and carbon dioxide in balance for the plants and animals to breathe was clearly the plan. The side of effect of the Great Oxygenating Event was mass extinction of creatures that could not survive in the new conditions.

But was it really extinction? Was it not obsolescence? Those creatures had done their job, and now Life was moving on. I remember going to Recycling Centres several years ago and there were always piles of old TV's. It was a shame because they would have all worked, but along had come new flat screen technology, which had made these TV's obsolete. Replacing old with new is not extinction.

However, cutting down rainforests, or poaching elephants for their ivory, just destroying without any replacement, that causes extinction.

There was far more oxygen in the atmosphere at that time, but it would have been necessary to go high before introducing vast numbers of oxygen breathing creatures that would bring the levels down, and in balance with the carbon dioxide. The high levels of oxygen are believed to have contributed to super-sizing of animals; the dinosaurs of course, and there was a dragonfly with a 70cm wingspan, and a 2.5m long centipede.

Shrinking Creatures
Dragonflies and centipedes have now become smaller with the lower percentage of oxygen. As we pour ever more CO_2 into the atmosphere, does this mean that we will get smaller too? An average man will go down to 80cm, a baby will be 8cm, while fruit and vegetables will supersize. 'Do I really have to carry the banana to school, Mummy?' 'Don't complain it is our turn this week.' 'But I took the tomato on Monday.' This could be an ingenious plan to fit more of us onto the planet without depleting its resources (or perhaps not).

We can also see how fish appeared earliest of all. Life that started in the oceans, protected from the harsher conditions, could move to each next stage quicker than life on land.

Tall trees that covered the Earth needed deep soil for their roots. The soil was made from trillions of dead organisms and had built up over millions of years. Flowering trees came at a later stage when there were also insects to pollinate them.

It seems amazing that insects and flowering plants should work together, but we must remember that Life is one thing with many separate forms. Life is on both sides; it is insects, and it is trees. The insects and the trees don't know what is going on, they just do what they are programmed to do to make this relationship work. At least 80 percent of plants are flowering plants making this the most successful design.

There would have been exciting phases in the evolution of Life when thousands of new and varied plants and creatures, never previously seen on Earth, were appearing. We have moved on from those phases now, so lifeforms that become extinct now will never be replaced.

But have you noticed something amazing? As Life became more complex and diverse, the planet became more beautiful. The early noxious atmosphere was tamed, the simple machine-like organisms became plants with beautiful flowers, tropical birds of paradise appeared, weird and wonderful insects, multi-coloured animals with their courtship rituals, and so on.

A picture of a Hawaiian island appeared on my computer screen, chosen for its beauty. This island would have started, not so long ago, as lava from a submerged volcano. Surprisingly quickly plants would have started growing on the black barren landscape, seeds blown by the wind or dropped by birds.

The island was transformed into a tropical paradise, with coral reefs, and shoals of colourful fish. It would have been a high-speed version of what Life did from the start.

Hawaii

2.SIX

Organic Machines

It is worth thinking again about little organic 'machines'.

The earliest ones had the job of purifying the atmosphere by reducing the carbon dioxide and increasing the oxygen. This was atmospheric clean-up on a global scale. They did this by taking in CO_2 and water and excreting oxygen to build their hydrocarbon structure.

There was 1000 times more methane in the early atmosphere; a potent greenhouse gas as we know. Methane is CH – carbon and hydrogen – which is a perfect building material for organisms made of hydrocarbons. It would make sense that there would have also been methane breathing creatures in this early World.

Of course, by burning fossil fuels, we have inadvertently reversed the clean-up process.

When we think about our own bodies, there is so much chemistry going on to create the materials that keep things functioning. For example, hydrochloric acid (HCl) is required by the stomach to digest our food. It seems a bit strange that we would ingest chlorine, but chlorine of course is found in salt (NaCl). This is partly why we need salt in our diets. Take hydrogen from water and chlorine from salt to turn salty water into acid. There are many other examples too, such as lubricants for our joints, and enzymes: catalysts for chemical reactions in the body.

Life created many organisms – and organs – to do this type of activity, but it also thought, if I have already created the functionality then why not reuse it? Some of the tasks can be sub-contracted to other organisms. This is why we have trillions of gut bacteria known as gut microbiota; all those adverts for probiotic yoghurt that feeds our good bacteria. Cows would not be able to digest grass without billions of bacteria in their stomachs.

These bacteria feed off our food and do the trick of excreting what we need to make our bodies function. They even work with us to tell our brains what to eat for the benefit of both them and us.

Wimbledon in the Future
Two women, one a lot older than the other, are looking at the menu in one of the cafes of the Wimbledon Grass Court Tennis Championships.

'We simply must have strawberries while we are here, it is the tradition,' the younger woman is saying as the young waiter approaches.

'Good afternoon ladies, have you decided?'

'We would like strawberries please,' says the younger woman, smiling at the young waiter.

'Would you like grass with your strawberries?' asks the waiter.

And seeing their undecided expressions, he continues, 'Five strawberries sprinkled with Centre Court grass is £50, Court One is £40, and grass from the other courts is £25'.

Looking at each other, the younger woman says, 'We are only here once, do you think we should go for Centre Court grass. It is rather expensive though - £100 for 10 strawberries with grass sprinkled on top?'

Before answering, the older woman turns to the waiter and says, 'I remember that a long time ago when I was a girl, they used to do strawberries with cream. You don't still do that do you?'

The young waiter looks puzzled and asks, 'What is cream?'

The older woman responds, 'Instead of serving the grass directly on the strawberries, the grass is first eaten by a cow.'

As it is evident that the waiter hasn't understood at all, she continues in her slow, patient manner perfected over years of teaching disinterested teenagers, 'The grass is chewed around in the cow's mouth with saliva before being swallowed and regurgitated a few times.

'Then bacteria in the cow's intestine break it down, and a bit more processing goes on inside the cow's body until white milk can be sucked out of the cow's udders, and from this you can make cream. Normally you whisk the cream to make it thick and put a dollop of it on the strawberries. I remember it being extremely delicious.'

The young waiter looks at the older woman assuming she must have dementia and in a slightly patronising voice says, 'No madam, I am afraid we only have normal grass, not grass that has first been eaten by a cow.'

2.SEVEN

The First Creatures

Let us continue to think about how Life might have carried out its plan to become more complex and more diverse.

Have you ever considered how similar an egg and a seed are? We had blackfly eggs on the runner beans, tiny black eggs. They look almost identical to the poppy seeds we use on homemade bread and pastries. The big difference of course is that one will become a fly and the other a flower.

Blackfly Eggs and Poppy Seeds

If you take a seed, maybe think of a bigger one like courgette, or cucumber, and put it in moist soil or water, it will germinate. The moist environment triggers it into action. Out of the top comes a little white stalk with two baby leaves. Meanwhile another white stalk with baby roots comes out of the other end. The top end starts photosynthesising and the roots go in search of nutrients. There were just enough nutrients – protein, minerals, vitamins – in the seed to get the process started. That is why seeds are a good source of food for us too.

Now consider an egg – let us think of a normal chicken egg – a fertilised egg does much the same thing. The DNA is in a whitish spot on the yolk, the yolk is made of fats, proteins, and essential nutrients – all the building materials required to make a chick. The egg white protects the chick and provides additional food before it hatches.

These are quite sophisticated seeds and eggs, but in a simpler form, it must be possible for eggs and seeds to be almost identical. The DNA tells the organism what to make and the nutrients provide the building materials for the baby organism.

There was a time when there were no oxygen-breathing creatures, there were just plants. Life needed to make the jump to create creatures that could move around. Plants had to give birth to animals, most likely initially to insects. The DNA inside a seed must have been changed to make it an egg from which an insect would hatch. How else could this happen? And yet it is strange for us to think that we evolved from plants.

As with our earlier discussion, Life has its own lifecycle, and each individual organism has theirs.

The plan to go to the next stage, along with the necessary code had already been prepared.

Tree Fly
The TV presenter is crouched down beside a small tree and is speaking in soft tones familiar to watchers of BBC wildlife programmes. He is pointing at a small fly.

'And here we have the 'tree fly,' he is telling the camera. 'She is laying her egg in the ground as you can see.' And continuing, 'The egg is really a seed from which a small tree exactly like this one will grow.

'The tree will develop seeds, which are really eggs, from which the larvae will hatch. The larvae will then feed on the leaves of the tree, and thus grow into adult flies. And the whole cycle will repeat itself.'

A lifecycle like this might be possible, though a bit pointless, best just do the jump from seed to egg.

Of course, it means that we also evolved from insects. Next time we feel like squashing a fly we could stop and think, if evolution had followed a different path, that could have been me.

Chicken or Egg
Which came first, the chicken or the egg? The egg! The egg came first with the DNA and nutrients to make the first chicks.

2.EIGHT
Evolution

When a mother gives birth to a child, people will say, 'Oooh, doesn't he look just like his Dad!' Imagine the shock, if a completely different creature popped out instead, but that must have happened at different stages in Life's lifecycle.

Turtle Skeleton

There was an article that pointed out that a turtle's skeleton was the same as other animal skeletons, except that the ribs had fused together to form the shell. Take a look at a turtle skeleton and you can see what they mean. The article continued to say that it must have evolved from another animal with a normal skeleton. What animal would it have evolved from? A camel maybe?

Camel to Turtle

The female camel was standing, chewing with her jaw making a circular motion which did not make her look that attractive when Life spoke to her. 'You have been chosen to start a new species which will be called a turtle.'

'Excuse me. What on Earth are you going on about?' replied the camel.

'Let me explain,' said Life. 'To create a new species, we must evolve it from an existing one, and in this case, we need to evolve it from a species with a skeleton like yours. Obviously with your hump, it gives us a head start with the shell. That is why you have been specially chosen.'

'But I don't want to be specially chosen,' said the camel. 'I want to have normal camel babies like all the other camel mums.' And at this point she had a tantrum with hind legs kicking out in a dangerous way.

Some while later she gave birth, and despite her baby looking rather strange, being as it was part camel and part turtle, she loved him very much.

It was therefore so sad that when he grew up and they made him crawl down the beach into the sea, like turtles do, that – despite heroic efforts to swim – with his leaky ribs and camel hind legs, he sank to the bottom and drowned.

This is an important point. People talk about one thing evolving from another thing, they are excited when they find a dinosaur fossil with feathers to provide the missing link between dinosaurs and birds, but something that is half one thing and half something else may be useless and unable to survive.

We see the same in the products that we create. Bicycles existed in a basic form for decades. It was such a brilliantly simple design, and so useful, that millions were made. All of China rode around on identical black 'Racing Pigeon' bicycles at one time. That good, solid design subsequently evolved into myriads of different versions. This would never have happened if the first design were useless.

Each new species that is created must be fully functioning, able to survive and reproduce, before it can evolve into different versions of itself.

Vacuum Cleaner

A woman is holding on the phone waiting to be put through to the service department of her vacuum cleaner company, having been assured many times that her call is extremely important to them.

Suddenly, the voice of a service engineer cuts off the annoying music to enquire, 'Can I help you?'

'Yes,' says the woman. 'I have a problem with my vacuum cleaner.'

'What exactly seems to be the problem madam?' enquires the service engineer in as concerned sounding voice as he can muster.

'Well, my husband was helping me vacuum the stairs. You see the machine is too heavy and awkward for me, so he was helping when he suddenly let out a scream, started rolling around in agony saying that he had ruptured himself, while the machine banged down the stairs and smashed a huge hole in the wall at the bottom.'

'Oh, I see,' says the service engineer. 'Which model is your vacuum cleaner madam?'

'It is the Vac-wash 50,' she replies.

'Ahh,' says the service engineering knowingly. 'As you know when we bring out a new product, we have to evolve it from an existing product. That is how it works. In this case we decided to evolve our new vacuum cleaner from our existing washing machine. The Vac-wash 50 is half vacuum cleaner and half washing machine. That is why it is a huge white metal box with half of ton of concrete inside it.'

The engineer continues, 'The scenario you describe is not totally unexpected, because, in its current form, the machine is rather useless'.

'Right,' says the woman uncertainly. 'May I ask when you plan to make some improvements?'

'Well, it doesn't really work like that,' says the engineer. 'But what I can say is, based on previous experience, the Vac-wash 75 will have probably evolved by some time next year, and we will get the full Vac 100 the year after.

'I am so glad to have been of assistance to you. Thank you for your valued custom. Goodbye.'

There are two things going on. There are big changes when, for example, a plant gives birth to an insect. Then there are smaller changes in which an existing organism evolves to better be able to survive, potentially evolving so much that it becomes a different species.

The big change is part of the lifecycle of Life itself in its journey to become more complex and diverse. The small changes are done at local level and are part of the lifecycle and evolution of individual species.

The big change involves a completely updated set of DNA instructions, whereas the small change involves a simpler modification. For example, the Peppered Moth changed from white to black to hide itself when our cities were polluted from burning coal. It then turned back to white again when the cities were cleaned up.

There is a mechanism for changing the instructions in the child. However, there are limitations on how big a change can be made between parent and child, without causing too much shock to the parent, or too extreme a difference in the child.

Going from seed to egg must be quite simple. Eggs are good because different creatures can hatch from them without any need for the parent to do anything. Think of our baby turtles hatching out after the mother turtle has long gone and carrying out their instructions to head to the sea. Also note how fish hatch from eggs, and algae from spores, so there can be a similar thing going on with the evolution of sea creatures.

Going from an egg to a mammal is more of a challenge. We can imagine female reptiles having a pelvis big enough to push out a big egg, so a modified version could push out a live baby. How would this stage work? A reptile would produce an egg, but the DNA would have been changed so that a mammal version of the parent would hatch from the egg. This would be a big change involving changes to behaviour too, because mammals need the necessary parental instincts to care for the child. Programmed behaviour is part of the genetic code.

Life did not have to do this, but it was all part of its plan to become more complex, and live births allow the most complex organisms to be created.

The new creature must already be adapted to its environment, but Life is on the ground in the creatures and so Life intimately understands the local environment. As the environment changes, Life must then be prepared to adapt the design to ensure its survival.

Darwin was troubled by what he called the 'abominable problem' that flowering trees appeared on Earth in a short timeframe and with great diversity. This should not have happened according to his theory of evolution. Whilst we owe him a great deal through his incredible insights, it does seem that he missed the point that Life itself has its lifecycle. If Life is ready to introduce the flowering trees together with the insects that do the pollination, then it can. It does not have to wait for a random mutation to achieve this.

Galapagos

Life was feeling rather contented with itself. I am good, thought Life. I have landed myself a really cushy number here on this beautiful Galapagos island. Warm air. Beautiful Pacific sea views. No other forms of me trying to eat me.

It could be a lot worse. I could have become a penguin battling the Antarctic winds, or a desert mouse hiding from the heat during the day, or... and the list went on.

What a master stroke to turn myself into a giant tortoise with a short neck. True, my tortoise friends are not the most exciting, or intelligent, but all in all I am good.

But something was troubling Life, something that the tortoises themselves were blissfully unaware of. There seemed to be an ever-decreasing amount of ground-level food to eat.

We are going to have to do something about it, or it will be the end of the line for this version of me. And that is unacceptable. Adapt and survive has been my moto for billions of years.

Right, thought Life, let's go for a tour of the island to check there are no hidden food supplies that I don't know about. And, sitting up in the cockpit of the tortoise's brain and looking out of its eyes, Life and tortoise go off on a journey.

This is so boring, despaired life. Cannot we go a bit faster? Oh, I remember, we are a tortoise, and tortoises are not built for speed. I will have to chill.

Eventually they completed their tour of the island. As expected, it was not good news. There is going to be a terminal shortage of food that we can reach with our short necks. What can we do?

The first thing is always to try to adapt our behaviour, like when we taught those monkeys how to break open nuts with stones.

OK, let's try to get my tortoise to climb on the back of that other one to see if we can reach some leaves hanging down from the trees. Oh dear. That was a disaster! What with our flat shiny front, and his shiny domed shell, we keep slipping off. This is never going to work.

Life pondered a bit, once we have gone down the route of a particular species design, our options become limited. However, we can always do a genetic change, an enhancement. Let's see, I can make some DNA changes to have a longer neck and modify the shell a bit around the neck area for more flexibility.

OK, I have the design, let's get to work with producing a new variant of me. We need to go and find a female to have sex with!

I don't care about you my short-necked friend, you are going to be toast anyway, thought Life, bouncing around a bit as the tortoises made love. I care about me and my continued survival into the next generation.

And when the tortoise turned its head, Life could see all the other turtles making love, because Life was in the brain of all of them, so they all had the same idea. No need to rely on some random mutation and rebuilding the population more-or-less from scratch when you can do the modification across the board.

Of course, the tortoises hadn't a clue what is going on. Who would? But they might have noticed that their children looked a bit different with their long necks, though maybe they weren't that observant.

And as the children grew up, they learned, with a bit of help from Life, how to eat the leaves from the trees. There was a plentiful supply of those.

Their parents couldn't reach unfortunately.

Is it not a bit unfair on Life to put these adaptations down to some mutant gene?

Yes, it is only the ones that have the mutation who survive, but that was the point, Life knew it needed to make the change. It did not need to sacrifice 90% of the population to do this.

I read how scientists took DNA from luminous coral and put it in fish DNA to make luminous fish for people's fish tanks. A scientist friend said to me, 'Oh that is easy. Anyone can do that.' Life would say, 'Making a small change to an existing species is easy. Anyone can do that! I have done much, much more difficult things over millions of years.'

2.NINE

DNA

It is worth contemplating DNA for a moment.

DNA is a molecule but, when a sperm fertilises an egg, it has been observed how his and her DNA molecules uncoil their double helices and do the thing where they take some from his and some from hers to create the new DNA for the child. This is the start of creating a new life.

Molecules don't do this; they don't have behaviour, not unless they are this amazing type of molecule. I remember a school chemistry class in which we mixed some chemicals and created a polymer that we could lift out of the beaker and wind around a pencil. That is the sort of thing that normal molecules do.

The DNA molecule has the intelligence to know that this is a life-creating moment. A similar intelligence is in evidence when a seed decides to germinate, or a bacterial spore decides it is safe to wake up.

Let's cast our minds back to the beginning when there was no life, and we proposed that life started in the atmosphere. Scientists have done experiments to find prospective 'organic' molecules that will build themselves given the right conditions, with enough success to prove the concept. This would have been the trigger that told 'DNA' to create itself for the first time from the surrounding gases.

Now we have been discussing the life equivalent of a software update. Life has the intelligence to know how to update the code to create an organism that is better adapted to survive.

When we consider the lifecycle of Life itself, the code change is a major update that creates a completely new organism.

The intelligent behaviour of DNA is at the heart of Life itself.

2.TEN

Creature Design

Skeleton Design

It is Planet Life's turn to deliver the inter-departmental talk to their colleagues in Universal Building. On the table at the front is a half-size human skeleton dangling from a string connected to a metal support that sprouts from a wooden base. This was one that they used for prototyping.

'This is a human skeleton,' announces the presenter slightly unnecessarily as most of the attendees could see what it was, and anyway there was a label saying 'Human Skeleton' written in large letters on the base.

The presenter touches an area of the skull which makes the forefinger of the right hand move up and down very subtly. Only the few seated nearest the front can see it and they all laugh in that unnecessarily hearty way that people do when they get the joke but nobody else does, leaving the rest to turn to their neighbours and ask them, 'Why are they laughing?', which of course they don't know either.

'First,' continues the presenter. 'I want to talk to you about the design of this skeleton. The skeleton's purpose is to support the structure of the human, but unlike a tree, for example, humans move around, so the skeleton needs joints between its bones.

'Everything must work together,' she says. 'That is why one team must design the whole skeleton, otherwise the left hand wouldn't know what the right hand was doing, as it were.'

She touches another area of the skull and this causes both the left hand and the right hand to move at the wrists and the skull to look down and to the left and right in what would have been a bewildered expression.

Her audience laugh because they could all see what was happening this time, and they began to think that this would be a fun presentation.

'The movements can be extremely complex. Take walking for example,' she says, touching another area of the skull which started both legs walking with the arms moving back and forth and the skull looking left and right as if saying to hello to people as it walked.

'Or running,' she says, suddenly putting the skeleton in a fast run, startling her audience as it seemed to be running away from some imaginary danger.

'See how everything moves together, so you couldn't just have one team designing a hand and another a leg. That would never work.'

'The skeleton moves because some muscles contract, while others extend.' And she switches the skeleton to a new mode; muscles which had been opaque and largely invisible to the audience, now shone blue when contracting and red when extending. She dimmed the lights, and with the skeleton back in walking mode, the audience could see how all the muscles were working together to create the movement. They liked seeing that.

'We have to design all the muscles for the whole skeleton as a single system. One team does this, working with the skeleton team.'

And she continues, 'To activate the muscles, signals are sent from the brain through another system called the nervous system. We have some pre-set modes like this walking mode, so the human just decides to walk, and the brain activates walking mode. Alternatively, the human can activate, for example, a single finger.' Now stopping the walk and doing the action she had done at the start. 'And,' she continues, 'Some movements are fully automatic without the human even knowing they are going on.

'The nerve team oversee creating all of this, obviously working closely with the skeleton and muscle teams. Look for example at how all the main nerves go down inside the spine. The skeleton team must design the bones so these will fit inside.

'Muscles need fuel,' she is saying. 'This is provided by the blood.' She now reaches behind the table that the skeleton is sitting on and picks up a large item. 'This is a module called the heart-lung module,' she announces, opening the rib cage and fitting it in, whilst slotting its breathing tube up into the skull's nose and mouth. 'The heart-lung team design this module.

'The heart-lung module connects to a system of arteries and blood vessels, a bit like the rivers and streams that you create. Every part of the body must be connected to this system.

'Of course, we have other modules and systems too, such as eyes, the digestive system, the urinary system, and so on. but that is enough for now.

'I hope you have enjoyed this overview of the design process. Let's have a break and enjoy my colleague's famous chocolate brownies in the shape of some of our new creatures!'

There are of course, two versions of the design, the male version and the female version. It is a clever design – as ever – that enables two versions to be created with a few tweaks here and there, and parts that are more developed in one or the other.

The hardware versions require matching gender software. A British comedian does a perfect impression of his wife putting on her tights and tottering on high heels. The audience – particularly the women – find it hugely amusing to see a large man doing such a perfect impression of female behaviour. When the gender hardware and software are not fully aligned, it can cause difficulties.

The choice of whether the female or male version is created is not random. In the population as a whole, there will be slightly more males created to account for the greater mortality rate. After a war this increases even more to replace the males who have been lost.

Many Creatures

After the break, the presenter is standing next to the old overhead projector and her audience can see that she has a large pile of acetate sheets that they presume will have pictures on them.

'Previously we talked about how we design a human. The human body uses our most sophisticated design. We are rather proud of this design and how it can be adapted to create many different creatures.

'I am going to show you some examples of where we use the same design in different ways,' she says switching on the projector so they can see a picture of a dinosaur skeleton.

'This is **Allosaurus**, 12m long and 5m tall, capable of running at 55 km/h. Can you recognise all the same components – spine, arms and legs, rib cage, head, feet, etc?'

Her audience were nodding.

Allosaurus

Now, putting up the next slide, 'This is the skeleton of the **African Pigmy Mouse**, weighing just 300 grams and 50 mm long. Can you see how it is the same design? Isn't that brilliant – whoops, sorry I shouldn't boast!'

She goes through her pile of pictures saying things like, 'See how the horse has four legs rather than legs and arms' or 'See how the orangutan has huge arms so it can swing from tree to tree', and, 'See how the arms of this bird have been converted into wings', 'Look at this sealion that has been adapted to live in the sea', 'This kangaroo is really good at jumping,' and so on.

Someone in the audience asks, 'Presumably the dinosaur came first and then you made the others from that one?'

'It is true that the dinosaur was created a long time before most of the others,' she replies. 'But the design had already been created and tested long before that. The dinosaur was built to our standard design. We adapt the design to suit the environment in which the creature must survive. The kangaroo was created specially to live in Australia for example.'

She points to a poster on the wall behind her head, the sort of poster that companies put up to remind their employees what to do. It had a picture of a fly with big eyes on it, and she reads out the caption, which says: *Don't create an eye on the fly! Create a fly with a fully working eye.*

'Everything we do must be fully functioning and tested out before it is born. Can you imagine how long a blind fly would survive for? Certainly not long enough to try and find a partner to mate with so that we can have another go at improving the design.'

It is interesting to consider how a design must be the whole of something. If you want a new kitchen, you don't just start by updating the far-left corner without considering the rest of the kitchen.

Some parts of the human design are self-contained modules. Surgeons can perform a heart-lung transplant or a kidney transplant. Other human parts go throughout the body. We see this with our own products that have pipes and wires going everywhere together with separate modules that perform specific functions.

We can design a lot of flexibility into our products. With Computer Aided Design, we could design one screwdriver and then put in parameters to automatically generate the rest of the set. That is a simple example, the structure of an aircraft wing can be generated from a complex set of parameters.

Think also about options. As cars travel down the factory assembly line, different versions are being built, different colours, different engines, the heated seat option, and so on.

It looks very much like the same thing is going on with Life. Life can customise its designs to adapt each lifeform to its environment. Sometimes the customisations may not work too well – feathered dinosaurs that cannot fly are an example – and other times it produces such a brilliant design that it goes on and on.

Neither should we forget the ability of our bodies and minds to adapt to changing conditions; the Nepalese Sherpas adapted to breathe less oxygen at altitude, those who can survive in extreme heat or cold, and so on. Of course, athletes use this feature to train their bodies to be better at specific athletic pursuits.

2.ELEVEN

Building a Human

Build a Human Session One

Planet Life continue with their presentations to Universal Building.

'Previously we talked about the design of the human body. Now we are going to look at building a human.'

'First I want to remind you of some basic principles,' she says, switching on their rather ancient overhead projector. A printed acetate sheet is already lying on the bed of the projector. This is now projected on the screen so everyone can see it.

'This is a single celled bacterium', she announces and lets her audience look at it for a short time before continuing. 'You will remember that we build everything in the same way. A single-celled organism like this, is a fully functioning living thing. Obviously, it cannot move around like a human or do other things that a human can do. But it can breathe in nutrients through its cell wall and excrete the waste products. It uses the atoms in the nutrients to build a copy of itself and then it will reproduce by splitting in two.

'As you see, basic behaviour of a living thing.

'This is all controlled by its DNA. The DNA tells it how to behave. It tells it how to breathe in the nutrients, how to make a copy of itself, and then how to split in two. And the copy it makes includes a copy of its DNA. It is an exact clone that will behave in the same way.'

Her audience are looking at the screen, but she can tell by their faces that they have understood so far.

'When we build a human, you must remember that the method is the same. Everything is created in the same way, because everything is part of one bigger thing called Life. The big difference is that the human is not one cell, but many cells that are joined together to create one much more complex organism.'

'How many cells?' asks one of the Universal Builders.

'About 37 trillion,' says the presenter in a matter-of-fact way, and adding some extra information that she thinks her audience will appreciate. 'These are made from seven million trillion atoms; 2/3 hydrogen, 1/4 oxygen, and about 1/10 is carbon plus some others.'

'OK. Not very many then,' replies the universal builder, who is used to working in much larger quantities.

'When we start building a human,' continues the presenter. 'We start with one cell that splits into two and then into four, and so on. Just like the bacterium, but these cells are building one human, not separate bacteria.'

'So, you start with one, then two, then four until you get to 37 trillion,' says a keen universal builder.

'Exactly!' she replies, and then thinks for a second. 'Well, give or take a couple of billion anyway.' And walking to the overhead projector she replaces the acetate sheet with a new one. This one has nine separate images.

'It takes nine months to build a human inside its mother. This is what it looks like after month one, month two, and so on up to month nine.'

'It goes from looking more like just a blob, through to looking like a mini human,' says one of her audience.

'Yes, it does. In the earliest images,' she says smiling, 'it is totally blobby because it is just cells without any clear definition.'

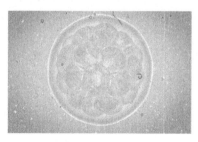

Human Embryo

'What you can see from the pictures,' she continues, 'is that we build in phases.' And looking at the questioning faces of her audience, 'We build the whole human at once.

We don't start just with a foot for example. And, as you just said, the whole human goes from more blobby to more recognisably human.

'When we plan the build, we decide what to do in each phase. The first phase is triggered when the sperm fertilises the egg, the second phase is triggered when the first phase has finished, and so on. This is why the whole process always takes roughly the same amount of time.

'Within a phase we decide on the design of each part of the human, as well as the sequence of steps to make that part. The design may be different for each phase.'

One of the audience says, 'I can see how the ears look different at the end of each phase.'

'Yes. That is a good example of parts being created to intermediate designs before the full design is reached', she replies. 'Actually, in one of our prototypes, we went for full grandad-sized ears as well as a big nose at eight months, but that caused some problems, so we went back to much smaller nose and ears in the final baby design.'

'One thing to note,' she continued. 'All the steps in a phase must complete before the next phase starts. If the leg building for a phase is complete, but the internal organs are still being built for that phase, then leg building stops and waits for the other tasks in that phase to catch up.

'These building phases continue after the baby is born and the child has grown up.

'Well, that is an introduction to how we build a human. Let's have another break, and we will go into some more detail in the next session.'

Pregnancies take the same amount of time because that is how long it takes to run through all the phases. It would be confusing if it wasn't like that.

Pregnant

Two young women walk up to each other, one heavily pregnant holding the small of her back, the other pushing a pram.

'Oh, you have had yours already,' exclaims the pregnant one.

'Yes, I was lucky, it only took three months.'

'Massively premature then,' says the pregnant one.

'No full term, he was 3 kg. What about you?' comes the reply.

'You are so lucky; I have been pregnant for 15 months. The doctor says it will be coming soon, but I don't believe she has any idea.'

Some creatures can modify the time taken to go through the phases. The unhatched chicks of some birds communicate with each other to synchronise their development. They will all hatch at the same time even though the eggs were laid over a two-week period.

Instead of one part of the body waiting for another to catch up, the chicks that are most developed wait for the others to catch up.

Build a Human Session Two

When everyone has returned from their break and settled down, the presenter continues.

'Remember that DNA contains the code for building an organism,' she says. 'If the organism has just one cell then it contains the code just for that one cell.'

'And,' interjects one of the universal builders triumphantly, 'if it contains 37 trillion cells, then the DNA contains the code for all 37 trillion!'

'That is right, but we are rather good – though I say it myself – at avoiding duplicate code. We only need one chunk of code to manage a blood cell for example because blood cells are all the same – and there are 25 trillion of those – or a skin cell and there are 35 billion of those. That cuts it down somewhat.'

'Now, an important bit,' she continues. 'When we create an organism, we don't just build it in phases, but also in a hierarchical structure. Think of an artist sketching out the rough layout for a picture. We do the same in three dimensions.

'We define that the organism will be split, say, into 60 large blobby sections. The first cell splits into 60, one cell for each section, with each of these cells positioned in the right place for the bit they will finally become. Each cell has a copy of the DNA, but we already tag the DNA so that it is specific to that section.

'Then we repeat this for the next level of the hierarchy. Each of these 60 sections know what they will become, and therefore know how many cells they need to split into for this next level down. This will be different for different sections. They split again into cells that will become the next level sub-sections.

'We go on like this down the structure until we have our framework. The more complex the organism, the more levels there will be in the hierarchy.

'Some of our framework cells will become a whole thing like a kidney, but others may be a mixture like the junction of where a bone and a muscle attach.

Blood Vessels

'Once the structure is complete, cells for blood vessels, or bones or tendons or whatever can be created, by running the sequence of steps for that tiny section in that phase. We can build a tiny section of blood vessel, for example, that will join up with other blood vessels being built at the same time, to become a complete network of veins and arteries and blood vessels according to the overall design.

'We also use the hierarchical structure to communicate back up. This allows us to ensure that all tasks for a phase are complete before we trigger the next phase.

'So,' asks one of the audience. 'Each cell has the complete DNA, but only activates their specific bit?'

'Yes. A bone cell in the knee contains the code for building a brain cell. However, by tagging the section that it belongs to, we narrow down the code that it is able to run.'

These insights come from understanding how we create our most complex products.

At its peak, there can be tens of thousands of people working to create an aircraft. How do all their separate bits build into the whole? It is down to how the product was split up in the first place through its product structure.

The product is split into major sections, then each of these is split again, down and down until the level where individual parts are created. It must be the same in a multi-cellular organism. In the early stages it will look like blobby cells randomly splitting, but it will not be random. It will split down in a product structure, both functionally and positionally.

The level will be reached where an individual cell knows – for example – that it is to become the wall of a tiny blood capillary. The cells around will be doing the same, and eventually everything will join up to create the whole vascular system. How else could it be done?

2. TWELVE

Our Body Machine

When we pause to think about it, we have been supplied with the most incredible 'body machine'. We are aware of our own breathing and our heart beating, but we don't really know what is going on inside. Our bodies tell us when they are hungry or thirsty, when they are too hot or cold, when they are tired, when they need to go to the toilet, and so on. They are working away all the time, repairing themselves, and fighting off infections.

Life is at work too, ensuring its survival and encouraging us to create and look after the next generation.

We alone amongst the lifeforms have the additional bit that makes us human. We have free will to do our extra things and make our own decisions. Incidentally, scientists have identified that the 'human brain' is in the front part of the brain.

We are all different with different interests, different skills, and different personalities. How much of this was pre-determined by our genes?

Selfo Setting

The Planet Life team were hard at work on their test planet, working on default settings for humans.

They were trying to decide on the 'Selfo' setting. Selfishness or self-centred behaviour is measured on the Selfo scale, where zero Selfos is completely selfless and 100 Selfos is a complete megalomaniac.

They set everyone in the first community to just 10 Selfos making them extremely selfless. This is what they found.

Esra phoned her friend Jane, 'I found a lovely homemade apple pie outside my door, Jane.'

'That is good,' said Jane. 'I thought I would make it for you.'

And Esra thought, Phew! Thank goodness I got it right that Jane made it, otherwise it would have been terrible. Jane would have felt guilty that she hadn't made one, and then I would have felt guilty for mentioning it. But she didn't even leave a note.

And she replied to Jane, 'Oh that is so kind of you, but I really don't deserve it at all, and you didn't leave a note with it, so I didn't know it was you.'

'Of course, you deserve it Esra. All the things that you do for other people all the time. I didn't leave a note because it would have seemed like I wanted credit for it, which I don't.'

'I am so sorry that I wasn't in when you came, Jane,' said Esra.

'But you didn't know I was coming, as I didn't tell you.'

'I know, but maybe I should have stayed in, even though I wasn't expecting you, just in case you came anyway.'

'Don't worry about it Esra. Did you go anywhere nice?'

'Actually, I was driving around to yours to deliver an apple pie that I had made. I hope you found it. I didn't leave a note because I didn't want any credit for it.'

'That is so kind of you Esra,' said Jane. 'I found a delicious apple pie, but I was worried about mentioning it in case it was not you who made it, but I really don't deserve it.'

'But you do Jane. All the things you do for everybody. I thought it would be a nice little thank you.

'Actually,' continued Esra. 'I might have been back in time, but I was delayed on my way back.'

'Oh yes?' asked Jane.

'Yes, as I was driving, I suddenly saw a hedgehog halfway across the road, so I stopped. It then went into a ball, so I got out to pick it up and move it, but a man driving the other way also stopped because he had seen the hedgehog too. I could see that he really wanted to move the hedgehog, so I said, 'You can move the hedgehog if you like.' But he replied, 'No I couldn't possibly do that because you were here first, and I can see that you want to move it'.

'Now there was a queue building up in both directions. One man behind me said, 'I am on my way to very important business meeting, but anyway, if I miss the most important meeting in a decade because of a rolled-up hedgehog in the road, I am sure they will completely understand'.

'Would you like to move the hedgehog? I asked him, but he insisted that it shouldn't be him because someone else deserved to move it more than him. It went on a bit like this until the hedgehog undid itself from its ball and walked off the road, and we all thought, that was lucky.'

Meanwhile they set a second community to a high Selfo level making them extremely self-centred. They didn't want to risk setting it too high because everyone might kill each other, so they decided on 75 Selfos.

The Planet Life researcher walks up to a man who he has spotted in the central square.

'Excuse me,' he asks. 'Are you in charge here?'

'Yes,' smiles the man. 'I am in charge. What can I do for you?'

However, another man, standing nearby, was listening and he felt a need to approach. 'Actually, I am in charge here,' he announces.

'No! I am in charge!' repeats the first man forcibly.

With the noise levels rising, a couple of other men arrive and say completely in unison, 'I am in charge. What is going on?'

Soon there is a crowd who are all saying they are in charge. At this point another man approaches and nods to the researcher, 'I am in charge here, so I will sort this out.'

He puts his hand in the air and announces, 'Will you all be quiet please?' But all the others are also putting their hands in the air and saying, 'Will you all be quiet please'. So, nobody was being quiet.

Fearing that this was getting out of hand, the researcher, shouts over the noise, 'Where can I get a cup of tea around here?'

At this point the entire crowd stops trying to be in charge, and they all point in the same direction down the hill. They all say at once, 'Go down the hill there and you will find a café on the left-hand side.'

'Thank you!' says the researcher heading off down the hill.

Urrm, thought the Planet Life team, a low Selfo level would have appeared to be ideal because everyone is so nice and caring of each other, but unfortunately it seemed very unlikely that anything would get done. Meanwhile a high Selfo setting was useless if everyone was so self-centred.

It was clear that the Selfo setting had to be randomly set at birth using an algorithm, and this worked well, particularly for democratic government.

The low Selfo people had no interest in being in charge, they were just grateful if someone else would do it, and, as they always saw the best in people, they would believe everything that a would-be leader promised. Meanwhile the high Selfo people only wanted to be in charge, and would promise anything that would get them elected, whilst having no guilt if they then did something completely different.

2.THIRTEEN
Love and Joy

We discussed how the universe design was perfect. A scientist is not going to discover how something works and then say, 'Actually that was not designed very well, it should have been done like this instead.' Universes are essentially engineering designs, where the best design is the one that does the job in the simplest way.

With Life we have also seen how a beautifully simple design concept makes it work, but here we witness something on another level. We see art in its purest form.

There is love and joy coming out of creation. We can find humour in our own strange behaviour because we are part of all of this. We find joy from relationships, from the beauty of nature, from the colours and smells that we have been given the ability to sense, from the sound of birdsong, or of running water, of trees, of the smells of fresh air, the feeling of the breeze on our faces, and the warmth of the sun.

We have also been given music and singing, dance, art, poetry, and literature. All these creative things that did not have to be there but are there because they are important to our creator. Don't you agree that these things elevate us, lift us up?

So much of our art is an emotional response to the beauty of creation. It can cause deep emotions in us, even a feeling of spirituality. Amongst so many other things, we can think of landscape paintings, and portraits.

Life had a plan from the beginning to become more interesting and more beautiful. That dream must have seemed a long way off at the start when there was a lifeless planet with poisonous gases in the atmosphere and volcanoes spuing out sulphurous smoke. However, each organism played its small part to ensure that the planet was a tiny bit better after it had gone, so that the full beauty could be enjoyed by generations, such as ourselves, that came along way into the future. We humans have played our part too in adding to the full tapestry of Life. Our different cultures and religions adapted to our environment, our art, our creations, our buildings, and our knowledge. Everything left for future generations to build upon. This has worked pretty well for most of the seven million years since we first arrived on the scene.

Final Thoughts

It is a tiny bit of a shame* that, although humans have been around for seven million years, in the last 50 years or so we have done quite a lot to trash the planet, from space junk down to plastics on the deep ocean floor.

*British irony alert!

But let's look on the bright side! This is just a minor side-effect of us being so extremely brilliant and, with this brilliance, there will come a time when we can jump on a spaceship and start again on another planet.

Then again, maybe we are not quite as brilliant as we think.

Planet Zob

It had been a long and tedious journey. Four couples had been selected to colonise GCL51, which was a planet chosen for being almost identical in size to planet Earth and a similar distance from a similar sized sun. It was therefore picked out as having a good chance of life.

Rachel and Steve's story was typical. With house prices so high they had no chance of getting on the property ladder, and even applying for a job as bar staff there were over 10,000 applicants. And what with climate change and predictions of the end of the World, they thought they would be better off moving to another planet.

Ideally, an unmanned probe would have been sent to check it out first, but as it was 2 light years away and the spacecraft could only go at half the speed of light, that was a journey of 4 years. There really wasn't time or budget to make two journeys, so this was a journey into the unknown.

The couples had been assured that if the planet seemed unsuitable, the spacecraft could sling-shot around GCL51 and come back home again. However, it soon became apparent that there were only enough provisions for the four-year trip out. Mission Control said was an unfortunate oversight, but the couples weren't so sure seeing as all available storage space had been taken.

And now, after four years cooped up in a tin can, GCL51 – or Planet Zob as it was known to the locals – was coming into view.

'It is blue just like Planet Earth,' announced Rachel. 'That is promising'.

Their craft, created by a private space company, could switch from travelling 0.5C to slowing down into a low planet orbit, and even turn into aircraft mode so that it could land without a separate lander. And now they piloted it down to an orbit where the full beauty of the planet became apparent.

They could see snow covered mountains and beautiful rivers meandering across verdant pastures. There were trees with so many coloured leaves, and flowers that looked as if they were tended and cared for although they were growing wild.

'It is even more beautiful than Earth!' exclaimed Rachel.

'Maybe it is like Earth used to be in the past,' said Steve.

After orbiting a few times, they had selected an open field as a suitable landing site and the pilot switched to aircraft mode and started the descent.

Once on the ground, the full beauty became ever clearer to them. The weather was warm, the air pure to breathe with no need for breathing apparatus. The smells were beautiful. Gravity was the same as on Earth as predicted.

They hugged and high-fived each other. The emotion of years stuck inside that capsule, the uncertainty of knowing whether they would truly find another life supporting planet, and then the beauty that surpassed all their imaginations. It was all too much, and they all burst spontaneously into tears.

They planted a flag bearing the logo of the private company in whose craft they had been transported. As the flag drooped down, Steve rigged up a support wire to make it look like it was flapping in the wind. Years later, this would lead to conspiracy theories back on Earth as to whether they had ever truly landed on GCL51.

Next, they held an emotional ceremony including some uplifting singing, and they all proclaimed, 'We claim this land in the name of the leader of our space company.'

Then of course they did selfies and uploaded the pictures and sent messages that would arrive in two years' time back home, which they all found very amusing.

'This will be our home. We will continue the human race here on this pristine unspoilt planet,' they agreed enthusiastically.

And they lay down on the beautiful, lush grass and holding hands they closed their eyes. It was the first time in years that they had felt truly relaxed.

All the time, the group have been watched by a couple of Zobbians.

Ahh, that was so moving. It brought a tear to the eye. All their hopes for the future; for a new life, they were both thinking. Though when the aliens had started doing selfies, giggling, and sending messages home, sympathy levels had rather waned.

But let's face it, nobody is that keen on aliens landing on their planet, and Zobbians are slightly obsessive in how immaculate they like to keep their planet, so their main thoughts were, yuk, think of all those nasty germs, bacteria, viruses, and alien DNA which could make a real mess of our beautiful ecosystem.

So, at that moment, the Zobbians zapped the aliens with their Zob-Zapper 5.0's.

Wow! That was amazing, they were both thinking as they grinned at each other. What a bit of kit! It was the first time either of them had used the 5.0, which was a big improvement on the 4.0, including automatic setting of the zap level and beam size.

The first one exclaimed to the other, 'Did you see that? I zapped that group of four aliens with a single zap on wide beam. Then I turned it on that stupid flag thing, it did a tiny 'pip', and the flag was zapped.'

'Let's go and zap the alien spaceship,' they agreed enthusiastically.

They sauntered over, and one went to the front and the other to the back. They aimed their Zob-Zapper 5.0's and started zapping.

'A-ma-zing!' they both cried out together, turning to each other grinning. They had completely zapped off the front and back ends of the spacecraft. Not like just burning a large hole, entire sections had disappeared.

Their beams moved towards the middle, and with two final zaps they had finished. 'Eight zaps of the 5.0 to completely annihilate an alien spaceship. Incredible!' they said, congratulating each other.

Zapping turns everything back into atoms, elements like carbon, aluminium, copper, magnesium and whatever else aliens and their stuff are made off. All planets are made of the same atoms, so this renders all the potentially hazardous biological material harmless. The field was now covered in piles of different powders all mixed up.

Their mate arrived driving the Zob-o-Clean 200. This looked like a large ride-on lawnmower except it was floating above the ground. The driver grinned and waved at them, looking very content. He was thinking, when you absolutely love cleaning and having everything neat and tidy, to be given a Zob-o-Clean 200 is a dream.

He drove up and down the whole area, much as you do when mowing the grass, sucking up the piles of powder.

The Zob-o-Clean 200 sucked up 100% of the powder but managed not to suck up the grass. With typical Zobbian attention to detail, where tiny amounts of grass or dust had been picked up, they were ejected out of the back, but there was hardly any of that.

The powder was automatically analysed and separated into different elements. What looked like large jam jars were being filled with different coloured powders, and when a jar was full, a coloured label was printed in beautiful Zobbian script to identify its contents along with a picture. The picture for carbon was a pile of black dust identical to the contents of the jar. This was not up to Zobbian standards, but they could not think of a better image.

Once a jar was filled, the lid was automatically screwed on, and a red and white checked piece of paper was fixed over the lid with a large elastic band, just to make it look nicer.

'Good job team!' they congratulated each other. 'No remaining sign of aliens apart from some skid marks from the spaceship's wheels. Let's have some lunch and then come back and to re-turf those areas.'

'We love our planet and love keeping it so beautiful. What better way to spend a morning than zapping aliens and tidying up the mess? I am going to really enjoy my lunch today,' pronounced one of the Zobbians.

Printed in Great Britain
by Amazon